INTRODUCTION

The following interpretation of words, phrases and notions occurring in the text, and also biographical sketches which the scope and purpose of the book itself make it impracticable to elaborate, are appended with the view to facilitating its perusal.

AT-ONE-MENT (state of unity, unitariness); denotes the ultimate state of oneness towards which all evolutionary movement tends; applied to consciousness, indicates the final expansion of consciousness wherein it coincides with the universal consciousness in extent and quality of comprehension. As applied to things, denotes the unification of all movements, tendencies, and evolutions as a singularity; the end of all evolutionary activity (vide p. 270).

BELTRAMI, EUGENIO, was born at Cremona, November 16, 1835; there he attended the elementary schools, the gymnasium and the lyceum, excepting the scholastic year 1848-49 when he was at the Gymnasium of Venice, now known as Marco Polo. He finished his lyceal studies in the summer of 1853, and in the following autumn (November) became a student in the *Mathematical Faculty of the University of Pavia*, after having obtained a scholarship there on the Castiglioni Foundation in the *Collegio Ghisleri*.

In 1854, the succeeding year, he was expelled from this college in company with five of his colleagues who were accused of promoting "disorders" against the Abbot LEONARDI, rector of the college. The expulsion brought him many hardships and disappointments, and for two years he drifted along merely existing as his family was too poor to have him matriculated at another university. But in 1856, he went to Verona where he succeeded in securing employment as secretary to the engineer, DIDAY, in the

Government service of Lombardy-Venice. On January 10, 1857, he was dismissed from this position "for political reasons"; but as the annexation of Lombardy to Piedmont occurred soon thereafter, he became again attached to the office of DIDAY, his former employer, when it was transferred to Milan as a consequence of political changes.

At Milan BELTRAMI took up his mathematical education in real earnest as he now had access to Professor BRIOSCHI, his former tutor, and also LUIGI CREMONA. Through the influence of these two men he was designated (October 18, 1862) "Professore straordinario" in the University of Bologna. His work on *Surfaces of Constant Negative Curvature*, as the pseudosphere, and his application of the expression given by LOBACHEVSKI (q.v.) for the angle of parallelism, very definitely secure for him a place among the foremost workers in the field of the non-Euclidean geometry. He postulated a theorem, known as *Beltrami's Theorem*, which he stated as follows: "The center of a circle circumscribing a triangle is the center of gravity of the centers of its inscribed and escribed circles." He died in the year 1900. (Vide *Amer.Math.Mo.*, Vol. IX, p. 59.)

BOLYAI, JANOS (1802-1860), was born at Kleansenburg, Hungary. He is said to have inherited his mathematical genius from his father, BOLYAI FARKAS (1775-1856), who was born at Bolya, Hungary. Being a very spirited youth, his progress in his studies was most remarkable. He completed the curriculum at the Latin school when only twelve years of age. Was graduated from the Philosophical Curriculum as a result of two years of study and then entered the Viennese Academy of Engineers. Was appointed lieutenant at Temesvárlin, 1823, whence on November 3, 1823, he wrote his father: "I have discovered such magnificent things that I am myself astonished at them. It would be damage eternal if they were lost. When you see them, my father, you yourself will acknowledge it. Now, I cannot say more, only so much: that from nothing I have created another wholly new world." This letter was written in the Magyar language and has been preserved at the Marcos Vásárhely, Hungary. The mathematical conceptions formulated by him became the appendix of the *Tentamen*, a book which his father had written on the Theory of Parallels.

His *ScienceAbsoluteofSpace* was translated into the French in 1868 by the French mathematician, J. HOÜEL, to whom belongs the credit of popularizing the works both of Bolyai and Lobachevski. (Vide *Science*, n. s., Vol. 35, No. 906, 1912.)

CAYLEY, ARTHUR, born at Richmond, Surrey, England, August 16, 1821; studied at King's College school; entered Trinity College, Cambridge, already a well equipped mathematician at the age of seventeen. When but twenty-one years of age he took two of the highest honors in the University of Cambridge. He was Senior Wrangler and First Smith's Prizeman. He published his first paper in 1841 and this was followed by eight hundred memoirs.

For fourteen years he practiced as Conveyancer. In 1863 Lady Sadler's various trusts were consolidated, and a new Sadlerian professorship of Pure Mathematics was created for the express purpose of affording a place for Cayley. Meanwhile, as early as 1852 he was a fellow of the Royal Society; in 1858 he joined Sylvester and Stokes in publishing the *Quarterly JournalofPur eandAppliedMathematics* .

He was for a considerable time principal adviser as to the merits of all mathematical papers which were presented for publication to the Royal Society, the Astronomical Society, the Mathematical Society and the Cambridge Philosophical Society. He is said to have been the "most learned and erudite of mathematicians," and much of the material, therefore, which now constitutes the basis of the non-Euclidean geometry is due to his laborious efforts and comprehensive knowledge of mathematics. (Vide *ReviewofReviews* , Vol. II, 1895, Sketch, reprinted from *Monist*.)

CHAOGENY (Gr. Chaos, disorder, geny—generating, evolution); the evolution of chaos into order. A kosmic process involving the elaboration of the original, formless world-plasm into the first faint signs of orderliness; the beginnings of the movement of life or the Creative Logos in preparation of the field of evolution.

CHAOMORPHOGENY (Gr. Chaos, disorder—Morphe, form—geny, becoming, generating); evolution of the space-form, the universe; the establishment of the metes and bounds of the universe; also, the origination and characterization of all forms as to tendence, purpose and limitations.

CONCEPTUALIZATION—The act of conceptualizing, the formulation of concepts; the process by which the Thinker arrives at concepts; the logical procedure by which the consistency of a scheme of thought is established.

CONSTRUCTION, IDEAL—A purely formal conception; a theory, hypothesis; a logical determination not necessarily based upon facts, but possessing virtue because of consistency; a self-consistent scheme of thought.

COSMOS—Whenever the term "cosmos" appears in the text spelled as here shown, it refers to phenomena pertaining to the earth or the solar system; when spelled "kosmos" reference is made to the universe as a whole.

CRITERION OF TRUTH—Defined in the text as a four-fold standard of reference, embracing the following elements, namely, the causal, the sustentative, relational and developmental. Lacking any one of these, no view of truth is more than fragmentary. Applied to space, it contemplates an inquiry into the genesis or causal aspect, an accounting for the duration aspect, a recognition of its relation to the totality of objects, and lastly, a prophecy of its telesic or perfective culmination. This test has been applied to the study of space as sketched in the text and the conclusions reached are an outcome of the inquiry directed along these lines.

CURVATURE OF SPACE—A doctrine formulated by RIEMANN and which maintains that space is curved, and consequently, all lines drawn therein are curved lines. Professor PICKERING aptly describes the results of movements in a curved space by pointing out that if we go far enough east we arrive at the west; north, we arrive at the south; towards the zenith, we arrive at the nadir, and *viceversa* .

DEIFORM—The basic idea indicated is that the universe is the form or body of the supreme deity, since He is not only immanent in the kosmos, but sustains it by His life; that in order to create a manifested universe, it was necessary to limit, or sacrifice, in a measure, His own illimitability. Viewed in this light, the kosmos assumes an added significance.

DIMENSION—(L. Dimensio, to measure), measurement; a system of space measurement. The Euclidean geometry recognizes three dimensions or coördinates as being necessary to establish a point position; witness, the corners of a cube to form which three of the edges come together at a point. These edges represent coördinates. For the purposes of metageometry, the term dimension has been variously defined, as, direction, extent, a system of space measurement, or a system of coördinates. Regarded as a series of coördinates, it became possible to postulate a system which required four coördinates to estabish the position of a point, as in the hypercube. There may be five, six, seven, eight, or any number of such coördinate systems according to the kind of space involved in the calculations. Determinations based upon the logical necessities of the various coördinate systems have been found to be self-consistent throughout and, therefore, valid for metageometrical purposes. Much depends upon the definition; for, after the definition has been once determined it remains then merely to make inferences and conclusions conform to the intent of the definition.

DIVERSITY—Philosophically, the idea indicated has reference to all dissimilarities, differences, inequalities, divergent tendencies, movements and characteristics to be noted in the universum of life; the antithesis of kosmic unity; the natural outcome of life in seeking expression; the result of the fragmentative tendency of life.

DUODIM (duo, two; dim, abbreviation of dimension)—A hypothetical being supposed to be possessed with a consciousness adapted only to two dimensions; a dweller in "Flatland" or two-space whose scope of motility is limited to two directions, as on the face of a plane; a term invented by hyperspace advocates for the purpose of establishing by analogy some of the characteristics of the four space and also its rationality.

DUOPYKNON (duo, two—pyknon, primary unit in the process of kosmic involution, a condensation)—Secondary phase in the elaboration of chaos into kosmic order. DUOPYKNOSIS (duo, two, secondary—pyknosis, process of condensation and origination)—The second period of the involutionary movement of life during which the duadic plane of the kosmos is being established; the second, in the series of seven distinct phases, of space-genesis; dual differentiation of kosmic plasm. Duopyknosis contemplates that, in the passage of the kosmos-to-be from the plane of non-manifestation to the plane of manifestation, there are seven distinct, though interdependent and interrelational, stages through which life passes, and that, of these, it is the second. It relates to the plane of non-manifestation and is, therefore, beyond the ken of the intellectuality, being a symbol.

EGOPSYCHE (Ego, the self-conscious I—psyche, soul)—The mental, emotional and physical mechanisms of man, the Thinker. These include the purely mental system, the emotional or affective mechanism, the nerve-systems (cerebro-spinal and sympathetic) and the brain; the objective or sense-derived consciousness of the Thinker which is elaborated from the total mass of perceptions transmitted through the senses; the medium of self-consciousness; the intellectual consciousness as distinguished from the intuitional or omnipsychic consciousness (q.v.).

The notion of the egopsychic consciousness is based upon data already empirically determined from the mass of evidences everywhere observable. It seems to be apparent that there is a consciousness, a seat of knowledge, in man the content of which is unknown to the sense-consciousness. Dreams, premonitions, intuitions, impressions and the totality of all such phenomena substantiate this view. Furthermore, it is agreed that the source of the intuition is not identical with that of the intellect. The egopsychic consciousness, accordingly, is purely intellectual.

FLUXION, PSYCHIC—The difference between a mental image and an object; an image is the representation of certain salient or cardinal characteristics of an object, sufficient for identification; but an image is not congruent, in every respect, with the object. Thus when we perceive an object, although as BERGSON contends we perceive it in the place where it is and not in the brain, it is the image of the object which takes its place in memory and not the object itself. There is, of course, a marked disparity between this memory-image and the object. Even if the image possessed one of the properties of the object, as, size, it could not take its place in memory, and neither could it do so if it possessed any of the properties of the real object. Consciousness is such that all due allowances are made for these conditions and the mind is able to retain more or less exact knowledge of these properties in the image; but there is a difference, small though it may be. This difference is the psychic fluxional.

FOHAT (Skt)—A term applied to the Creative Logos who is said to be the generating element in the differentiation of chaos into kosmic orderliness;

the supreme deity in the rôle of Creator.

FORM, PURE—An abstraction arrived at by subtracting the last vestige of materiality or substantiality from an idea and viewing the remains as a pure unsubstantial form or idealization; the shell or frame-work of a material object or condition; existing in idea or thought only; a mental conception regarded as a type or norm; a purely hypothetical construction.

FOUR SPACE—Often referred to as the fourth dimension (vide Chapter V); a space in which four coördinates (four lines drawn perpendicular to each other) are necessary and sufficient to establish the position of a point, as, a hypercube.

GAUSS, CHARLES FREDERICK, born at Brunswick, April 30, 1777. His father, being a bricklayer, had intended that he should follow the same occupation. So, in 1784, Charles was sent to the Bütner Public School in Brunswick, in order that he might be taught the ordinary elements of education. But during his attendance at this school, his unusual intelligence and aptitude attracted the attention and friendship of Professor BARTELS who later became the Professor of Mathematics at Dorpat. In 1792, through the kindly representations of Professor BARTELS to the Duke of Brunswick, young Gauss was sent to the Collegium Carolinum. This greatly displeased his father as he saw in this move the frustration of his plans for Charles. In 1794, however, GAUSS entered the University of Göttingen still undecided whether he should make mathematics or philology his life work. While residing at Göttingen, he made his celebrated discoveries in analysis and these turned his attention definitely to the field of mathematics.

He completed his studies at Göttingen and returned to Brunswick in 1798, residing at Helmstadt where he had access to the Library in the preparation of his *Disquisitiones Arithmeticae* which was published in 1801. He received his doctorate degree (Ph.D.) on July 16, 1799. His next notable work was the invention of a method by which he calculated the elements of the orbit of the planet *Ceres* which had been discovered by PIAZZI, January 1, 1800, and who had left no record of his calculations by which other astronomers could locate the planet. GAUSS also calculated an ephemeris of *Ceres'* motion by means of which DE ZACH rediscovered the planet December 31, 1800.

His *Theoria Motus Corporum Coelestium in Conicis Sectionibus Solem Ambientium,* in which the author gives a "complete system of formulæ and processes for computing the movements of a heavenly body revolving in a conic section" is an outgrowth of his early researches and brought him lasting fame.

Through the influence of his friend OLBERS, he was appointed, July 9, 1807, first director of the new Göttingen Observatory, and Professor of Astronomy in the University, a position which he held until the end of his life. He died February 23, 1855. (Vide *Astronomical Society Notices*, Vol. 16, p. 80, 1856; also *Nature*, Vol. XV, pp. 533-537, 1877.)

GEOMETRISM—Of geometrical quality; a notion derived from PLATO's declaration: "God geometrizes." It was his belief that the creative acts of the deity are executed in accordance with geometric design and laws; that in the totality of such acts there necessarily inheres a latent geometric quality. KANT closely adhered to this notion in his discussions of space as an aspect of divine intelligence. He believed that the intellect merely rediscovers this latent geometrism when it turns to the study of materiality, and this belief is shared by BERGSON, the foremost metaphysician of the present time.

HYPERSPACE (hyper, above, beyond, transcending—space)—That species of space constructed by the intellect for convenience of measurement; an idealized construction; a purely arbitrary, conventional mathematical determination; the fourth dimension; any space that requires more than three coördinates to fix a point position in it, as, a five space, an *n*-space.

INTUITOGRAPH—The means by which the omnipsychic consciousness transmits intuitional impressions to the egopsychic or intellectual consciousness. An intuitogram is a direct cognition, an intuition; a primary truth projected into the egopsychic consciousness by the Thinker. It is recognized that, under the necessities of the present schematism of things, it is exceedingly difficult to propagate an intuition, especially with the same degree of ease as concepts are propagable; yet, this is believed to be a condition which will be overcome as the evolution of the higher faculties proceeds.

INVOLUTION—Process of enfolding, involving; antithesis of evolution; philosophically, the doctrine of involution maintains that, during the process of kosmic pyknosis (space-genesis), *all* that is to be expressed, developed and perfected as a result of the evolutionary movement was first involved, enfolded or deposited as latent archetypal tendencies and radicles in the original world-plasm; that, as the involutionary movement proceeded through the various phases of space-genesis, these became more and more phenomenal until at last they terminated in the elaboration of a manifest universe: each stage, accordingly, of the involutionary procedure became the basic substructure of a plane of specialized substance or materiality and consciousness. Thus it appears that evolution really begins where involution ends (vide Fig. 18), and the two opposing processes constitute the dualism of life as generating element. This notion has been symbolized in the *Lingam yoni* of Hellenistic philosophies, also in *Yang* and *Yin* of Chinese philosophy, which represent the original pair of opposites.

KATHEKOS—A purely arbitrary term devised for the express purpose of providing a convenient symbol to convey the idea embodied in the triglyph, *Chaos-Theos-Kosmos*, and is composed of the first three letters in each one of the terms of the triglyph; hence, symbolizes the triunity and interaction involved in the resolution of chaos into an orderly kosmos by the will of the Creative Logos. Thus, "kathekos" embodies a quadruplicate notion, namely, chaos, Creative Logos, manifested kosmos, and the creational activity of the Logos in the transmutation of disorder into order. The justification for this term, therefore, resides in its convenience, brevity and comprehensiveness.

By referring to figure 17, it will be seen that Kathekos divides into two kinds—involutionary, or that which pertains to involution, and evolutionary, or that that pertains to evolution. It thus comprises the beginning and the end of the world age or cycle and pertains to non-manifestation. The *raison d'être* of this differentiation is embodied in the notion that, on the involutionary arc of the cycle, the chaogenic period represents a phase of the world age when space-genesis is in an archetypal state wherein are involved all possibilities that are to become manifest in the kosmos, and on the evolutionary arc, the kathekotic period which is parallel to the chaogenic and represents a phase of the world age when the kosmos has reached ultimate perfection, embodying the perfected results of the possibilities which inhered in the chaogenic period or in involutionary kathekos. Thus, kathekos is dual in nature, on the one hand representing kosmic potency, and on the other, kosmic perfection of these potencies. It is Alpha, as related to involution, and Omega, as related to evolution.

KATHEKOSITY—A derivative, signifying creative activity and all that it implies; the state of consciousness or cognition corresponding thereto.

KLEIN, FELIX (1849—), born at Dusseldorf; studied at Bonn, and when only seventeen years of age was made assistant to the noted PLÜCKER in the Physical Institute. He took his doctorate degree in 1868, then went to Berlin, and later to Göttingen where he assisted in editing PLÜCKER'S works. He entered the Göttingen faculty in 1871; became Professor of Mathematics at Erlangen in 1872; and subsequently held professorships at Munich, 1875; Leipzig, 1880, and Göttingen, 1886. No one else in Germany has exerted so great influence upon American mathematics as he.

KOSMOS—*SeeCosmos* .

LA GRANGE, JOSEPH LOUIS, born at Turin, January 25, 1736; died at Paris, April 10, 1813; regarded as the greatest mathematician since the time of Newton. It may be interesting to note that LA GRANGE remarked that mechanics is really a branch of pure mathematics analogous to a geometry of four dimensions, namely, time, and the three coördinates of the point in space. (Vide *Ball'sAccountoftheHistoryofMathematics* .)

LIE, SOPHUS, a noted mathematician, referred to as the "great comparative anatomist of geometric theories, creator of the doctrines of Contact Transformations and Infinite Continuous Groups, and revolutionizer of the Theory of Differential Equations."

LOGOS—The supreme deity of the phenomenal universe; Creator; Fohat; a planetary god; the deity of a solar system.

MANVANTARA (Skt.)—A world age; the periods of involution and evolution combined; the stage during which the universe is in manifestation; a Day of Brahma.

MATHESIS (Gr. mathein, to learn)—Erudition; profound learning; the realm of metaphysical conceptions; the field of higher mathematics; the sphere of conceivability; the theoretical.

MENTOGRAPH—A cognitive factor consisting of a complete perception fused or in coalescence with a memory-image. Pure memory, of itself, is without utility as an aid to cognition; but, when nourished or supplemented by the substance of perception it becomes the basis of intellectual consciousness.

METAGEOMETRY (Gr. Meta, beyond, transcending—geometry)—Commonly, any kind of geometry that differs from the Euclidean, as the non-Euclidean; a geometry based upon the assumption that the angular sum of a triangle is greater or less than two right angles; the highest form of geometry; a system of idealized mathematical constructions. Sometimes called "pangeometry"; designated by GAUSS as "Astral Geometry"; the geometry of hyperspace. It consists of results arrived at by geometers in seeking a proof of the parallel-postulate.

META-SELF—The higher self in man; the universal self; the one self of which all individual selves are but fragments or parts. In man, it is coördinate with the omnipsyche (q.v.) and as such is the medium of kosmic consciousness.

MORPHOGENY (Gr. Morphe, form, vehicle, body—geny, evolution)—The evolution of forms, the production of individual bodies or vehicles for life, including organs and faculties. Morphogenic—a derivative; pertaining to morphogeny; a kosmic process (vide figs. 17 and 18).

N-DIMENSIONALITY—Quality of conceptual space by virtue of which it may be regarded as possessing an indefinite number of dimensions.

NEAR-TRUTH—Any statement or view which is based upon partial knowledge; predicates concerning a class or genus derived from limited acquaintance with particulars of the class or genus; statements based upon logical determinations inhering in idealized constructions and applied to concrete or objective conditions; an abstraction viewed as a reality; the application of the qualities of abstractions to realities.

NEUROGRAM—Psychologically, a movement received by the afferent nerves in the form of a stimulation and transmitted through the brain and efferent nerves as either a reflex or voluntary action; a nerve impulse; a perception; a primary unit of intellectual consciousness; cf. *Intuitogram*.

NEWCOMB, SIMON (1835-1909), born at Wallace, Nova Scotia; educated in his father's school and came to the United States in 1853. Began, in 1854, teaching in Maryland; was appointed computer on *Nautical Almanac* at Cambridge in 1857; was graduated at Lawrence Scientific School in 1858; appointed Professor of Mathematics in the U. S. Navy in 1861. He supervised the construction of the 26-inch equatorial telescope at the Naval Observatory, and was secretary of the Transit of Venus Commission; was a member of nearly all of the Imperial and Royal societies of Europe and of the various societies in the United States, receiving the Copley Medal in 1874; the Huygens, 1878; the Royal Society, 1890, and the Bruce Medal in 1898; held the presidency of the following learned societies, viz: American Association for the Advancement of Science, 1877; Society for Psychical Research, 1885-1886; American Mathematical Society, 1897-1898; the Astronomical and Astrophysical Society of America from its foundation in 1899. He rendered notable service in popularizing the doctrine of hyperspace.

NORM—An authoritative standard; model or type; standard of reference. The choice of a norm for spatial determinations cannot abide in any premise except that which naturally, and not artificially and conventionally, conforms to what is actually perceived; if so, there should be justification for challenging the wisdom and utility of the present schematism of things. There is an inherent conformity of space with intellect and intellect with space, and because of this natural complementarity of part with part and whole with whole, space cannot be otherwise than the intellectuality normally conceives it to be, provided, of course, that the cognitive movement is free and untrammeled by arbitrary hindrances. Consciousness, therefore, is the norm or standard of reference for all questions arising out of a consideration of spatiality.

OMNIPSYCHE—A term used to denote the Thinker's cognitive apparatus; the universal soul manifesting in individuals; the consciousness of the Thinker in virtue of which he is at-one with the universal consciousness; the medium of kosmic consciousness; the source of the intuition, cf. *Egopsyche*. The divinity in man (which is taken for granted), or his highest self can in no way be said justly to take its rise from sense-experience or from any bodily process. If divine, then eternal, and therefore, persistent. Broadly, the doctrine of evolution recognizes the passage of life from form to form, adding a little to each successive form and inevitably pushing each to a higher degree of perfection. Now, what is it that passes from form to form? Is it undifferentiated life or is it a specialized form of life? From every evidence, it would be judged that the life that ensouls an individual form is a specialized principle, i.e., limited to the execution of a given purpose. If life as a specialized principle, limited to the execution of a given purpose in each form, passes on, it must preserve, at least, the sublimated results obtained during its residence in each individual form. It would thus become a sort of reservoir containing all these transmuted results. The omnipsyche, within the meaning of the text, is precisely this specialized life principle.

PARALLEL-POSTULATE—Variously referred to as the XIth, XIIth and XIIIth axiom of the *Elements of Euclid*; stated by Manning, in *Non-EuclideanGeometry* , p. 91, in the following form: "If two lines are cut by a third, and the sum of the interior angles on the same side of the cutting line is less than two right angles, the lines will meet on that side when sufficiently produced." This celebrated postulate has proven to be the most fruitful ever devised; for it embodies in itself the possibility of three geometries based respectively upon the following assumptions, namely: I. That there exists a triangle, the sum of whose angles is congruent to a straight angle, the Euclidean; II. That there exists a triangle the sum of whose angles is less than a straight angle, the Lobachevskian; III. That there exists a triangle the sum of whose angles is greater than a straight angle, the Cayley-Klein. Speaking of the content of the last two named, Edward Moffat Weyer[1] says: "Hypothetical realms, wherein the dimensions of space are assumed to be greater in number than three, yield strange geometries, which are only card castles, products of a sort of intellectual play in the construction of which the laws of logic supply the rules of the game. The character of each is determined by whatsoever assumption its builder lays down at the start."

PASSAGE OF SPACE—A phrase connoting the movement of space from chaos to perfect order, a process believed to be infinite. The genesis of space necessarily implies an elaboration, a procedure, by which the metamorphosis from disorder to kosmic order is made, and this movement is referred to as the "passage of space," a phenomenon thought to be measurable by means of a suitable instrumentality.

PERISOPHISM—See *Near-Truth.*

PSEUDOSPHERE—A surface of constant negative curvature; basis of BELTRAMI's metageometrical calculations; surface resembling a champagne glass or common spool. The assumption that space is pseudospherical has given rise to the notion of space-curvature and various other conceptions.

PSYCHOGENY (Gr. Psyche, Soul—geny)—History of the evolution of the soul or the development of the senso-mechanism in organisms. ERNST HAECKEL has traced the psychogeny of man through twenty-two different stages from the moneron to the anthropoid apes, and man.

PRALAYA (Skt.)—Kosmic quietude; the period during which the universe is not in manifestation; gestatory period; kosmic inactivity; opposed to manvantara (q.v.); figuratively, the kosmic womb; world egg.

PYKNON (Gr. pyknon, hard)—The principle of kosmic condensation; the primary basis of space-genesis; the initiation of the process by virtue of which chaos is elaborated into kosmic order. PYKNOSIS—The process of spatial engenderment. There are seven of these processes, each indicating a phase of duration, namely: MONOPYKNOSIS, the primary phase; DUOPYKNOSIS, secondary; TRIPYKNOSIS, tertiary. These three pertain to the plane of non-manifestation, the pralayic or gestatory duration-phase. The results arrived at during these duration-phases are concentrated in the Quartopyknotic which corresponds to the causal plane of manifestation or pure kosmic spirituality. QUINTOPYKNOSIS, a process concerned in the genesis of mentality; SEXTOPYKNOSIS, kosmic sensibility; SEPTOPYKNOSIS, kosmic materiality. These seven phases of duration constitute the scope of space genesis or kosmogenesis, and incidentally depose the substructure of kosmic materiality, sensibility, intellectuality and spirituality, as well as the higher trinity of kosmic modes. The ramifications of these principles are innumerable and omnipresent. (See Chapter VII.)

QUARTODIM—A hypothetical being assumed to have a consciousness adapted to hyperspace or the fourth dimension, and whose scope of action is encompassed within a space which requires four coördinates, as the four-space.

REALITY (Realism)—Life; the harmony existing among the parts to maintain their equilibrium in the whole; the principle of integrity subsisting among parts; kosmic vitality.

RIEMANN, GEORGE FREDERICK BERNHARD, was born September 17, 1826, in the village of Breselenz, near Dannenburg, in Hanover. Until he was eight years of age his father was his sole tutor, but even at this age he exhibited great powers of arithmetical calculation. In the Spring of 1840 young RIEMANN was sent to the Hanover Lyceum where he remained for two years, leaving in 1842 for the Gymnasium at Luneburg. Here, under the direction of Professor SCHMALFUSS, he learned very rapidly, and is said to have required only one week thoroughly to familiarize himself with LEGENDRE'S *TheoryofNumbers* .

On April 12, 1846 (Easter), he entered the University of Göttingen as a student of Theology in accordance with his father's wishes. His passion for mathematics, however, was so aroused by the lectures of GAUSS that He begged his father to be allowed to devote himself entirely to the studies of his choice. For two years he studied under JACOBI at Berlin. He then returned to Göttingen, and was graduated, his thesis being a dissertation on the foundations of a general theory of functions of a variable complex magnitude. In 1854 he qualified as a teacher by giving a lecture on the "Hypothesis on which Geometry is Founded." In 1857 he became "Professor Extraordinarius," and in 1859 was elected Corresponding Member of the Academy of Sciences of Berlin and in 1860 a member of the Academy of Sciences of Göttingen.

After four years of failing health, during which he visited Messina, Palermo, Naples, Rome, Florence, Pisa and Milan, he died at Lago Maggiore, July 20, 1866, in full possession of his faculties and conscious of his approaching end.

SCHWEIKART, FERDINAND KARL (1780-1857), studied from 1796 to 1798 in Marburg, attending the mathematical lectures of J. K. F. HAUFF. In 1812 he became professor in Charkov, a position which he held for four years. In 1816 he became a tutor in the City of Marburg where he remained until 1820 when he transferred his labors to Königsberg. It was during his tutorship at Charkov, Marburg and Königsberg that he, entirely alone and without the slightest suggestion from any man, developed and taught a non-Euclidean geometry to the students under his care. For copy of his treatise on non-Euclidean geometry, see *Historical Sketch of the Hyperspace Movement*, Chapter II.

SCOPOGRAPHIC IMPRESSIONS—Sight perceptions fused with an associated memory-image, and forming the basis of action on external phenomena.

SENSOGRAPHIC IMPRESSIONS—Perceptions or impulses transmitted through the nerves of a sense-organ; any impression acting through the media of the senses.

SENSIBLE WORLD—The world of the senses; that which responds to the senses; the domain of perception; the phenomenal world; world of perceptual space.

SPACE-CURVATURE (see *CurvatureofSpace*).

SPACE-GENESIS—The process of spatial engenderment; the movement of life as engendering agent in bringing into manifestation the kosmos; the story of the appearance of the organized kosmos. The genesis of space can only be symbolized, as has been done in the text, for the limitations of human consciousness do not otherwise admit of the empirical establishment of the notion of its detailed procedure.

SPATIALITY—Space as a dynamic, creative movement; kosmic order, as opposed to disorder; the path of the engendering movement of life; the place of life. Spatiality, materiality, intellectuality and geometricity or the latent geometrism of the kosmos are thought of as being consubstantial and interdependent; but, of these, spatiality is regarded as the substance out of which the latter three are elaborated.

SUPERCONCEPTUAL—The purely intuitional; an act of cognition performed without the detailed work of conception derived from sense-data; conception of intuitions and their inter-relations; the Thinker's consciousness freed from intellectual characterization.

SUPERPERCEPTION—Perception of conceptual relations; a state of cognition wherein, instead of receiving percepts or images from the external world, then elaborating them into concepts, the Thinker apprehends composite images or concepts at first hand. It is a power which the liberated mind of the future will possess owing to the growing automatism of the intellect and the more facile expression of the intuitional consciousness.

TESSERACT (Gr. Tessera, four, cube, tessella)—A hypercube (see Chapter V.)

THINKER (Skt. Manu, thinker)—The real, spiritual man, as differentiated from his perceptive vehicles—mind, emotions and physical body; the omnipsychic intelligence who receives, classifies, interprets and preserves percepts; the manipulator of concepts; in fine, the higher, spiritual man.

The Thinker uses the various perceptive instrumentalities as so many tentacles or antennae by which he contacts the sensible world and makes the necessary adaptations to environment. He is the pure intelligence which is the source of all cognitive motivation; opposed to ego, because the egopsychic instrumentality is essentially an individualizing, separative agency; while the Thinker's omnipsychic intelligence is the basis of his unity with the universal intelligence. This conception of the Thinker implies that, as a spiritual intelligence, he is within and without the body, filling it as the ocean fills the sponge, encompassing, enveloping it and, at the same time, originating the totality of activities which manifest in and through the body. He is limited, therefore, in his manifestations in the sensible world only by the pliability of his vehicles.

TRANSFINITY—A state or condition that is incomprehensible to finite intelligence; that which transcends the finite, yet is not infinite; less than infinity and greater than finity. Space is referred to as being transfinite rather than infinite in extent. But space transfinite should be distinguished from space "finite though unbounded." For, there would seem to be little worthy of choice between a "finite, unbounded space" and an infinite one. The absence of boundary would naturally suggest an infinite extent. And although RIEMANN who is the author of the "unbounded" space arbitrarily determined that such a space should be a manifold possessing a measure of curvature which could be determined either by counting or actual measurement, he undoubtedly knew, nevertheless, that while each manifold might be an "unbounded" space the totality of such manifolds, infinite in number, must also be infinite in extent. It would seem to do violence to common sense, if not to logical necessity, to view space both as "unbounded" and finite in extent, yet there would be no such difficulty in the recognition of space as being both transfinite and finite; because it is conceivable that the extent and character of space finite should transcend a finite intellectuality, and yet not be infinite.

TRIDIM—A being whose scope of consciousness is limited to a space of three dimensions, as ordinary human beings. TRIDIMENSIONALITY—That quality possessed by perceptual space by virtue of which it is necessary and sufficient to have three coördinates, and only three, to establish the position of a point.

UNODIM—A hypothetical being assumed to have a consciousness limited to linear or one-space.

ZONES OF AFFINITY—Regions in the domain of intellectuality wherein minds, possessing a common differential, rate of vibration or quality, adhere to certain tenets from choice. Schools of philosophy, religions, and all those major divisions of intellectual effort which divide and subdivide intellectual allegiance are believed to take their rise in this property of intellectuality in virtue of which all minds having a similar coefficient gravitate towards a common agreement, especially where the movement is voluntary and untrammeled.

PART ONE

THE ESSENTIALS OF THE GEOMETRY OF HYPERSPACE AND THEIR SIGNIFICATIONS

CHAPTER I

On the Variability of Psychic Powers—The Discovery of the Fourth Dimension Marks a Distinct Stage in Psychogenesis—The Non-Methodical Character of Discoveries—The Three Periods of Psychogenetic Development—The Scope and Permissibility of Mathetic License—Kosmic Unitariness Underlying Diversity.

In presenting this volume to the public profound apologies are made to the professional mathematician for the temerity which is shown thereby. All technical discussion of the problems pertinent to the geometry of hyperspace, however, has been carefully avoided. The reader is, therefore, referred to the bibliography published at the end of this volume for matter relating to this aspect of the subject. The aim rather has been to outline briefly the progress of mathematical thought which has led up to the idea of the multiple dimensionality of space; to state the cardinal principles of the Non-Euclidean geometry and to offer an interpretation of the metageometrical concept in the light of the evolutionary nature of human faculties and material characteristics and properties.

The onus of this treatise is, therefore, to distinguish between what is commonly known as sensible space and that other species of space known as geometric spaces. Also to show that the notion which has been styled *hyperspace* is nothing more nor less than an evidence of the faint, early outcroppings in the human mind of a faculty which, in the course of time, will become the normal possession of the entire human race. Thus the weight of all presentations will be to give currency to the belief, very strongly held, that humanity, now in its infancy, is yet to evolve faculties and capabilities, both mental and spiritual, to a degree hitherto viewed as inconceivable.

On this view it must appear that the faculty of thought including the powers of imagination and conceptualization are not psychological invariants, but, on the other hand, are true variants. They are, consequently, answerable to the principle of evolution just as all vital phenomena are. Some have thought that no matter what idea may come into the mind of the human race or at what time the idea may be born the mind always has been able to conceive it. That is, many believe that the nature of mind is such that no matter how complex an idea may be there has always been in the mind the power of conceiving it. But this view cannot be said to have the support of any trustworthy testimony. If so, then the mind must at once be recognized as fully matured and capable during every epoch of human evolution, no less in the first than in the latest, which, of course, is absurd. It is undoubtedly more reasonable and correct to believe that the powers of conceptualization are matters of evolutionary concern. For instance, the assertion that the mind was incapable of conceiving, in the realm of theology, a non-anthropomorphic god, or, in the field of biology, the doctrine of evolution, or, in the domain of invention, the wireless telegraph, or, in mathematics, the concept of hyperspace before the actual time of these conceptions, cannot be successfully controverted.

In fact, it may be laid down as one of the first principles of psychogenesis that the mind rarely, if ever, conceives an idea until it has previously developed the power of conceptualizing it and giving it expression in the terms of prior experience. As in the growth of the body there are certain processes which require the full development of the organ of expression before they can be safely executed so in the phyletic development of faculties there are certain ideas, conceptions and scopes of mental vision which cannot be visualized or conceptualized until the basis for such mentation has been laid by the appearance of previously developed faculties of expression. And especially is this true of the intellect. Inasmuch as the entire content of the intellect is constituted of sense-derived knowledge, with the exception of intuitions which are not of intellectual origin though dependent upon the intellect for interpretation, there can be no doubt as to the necessity of there being first deposed in the intellect a sense-derived basis for intellection before it can become manifest. The Sensationalists, led by LEIBNITZ, propounded as their fundamental premise this dictum: "*There is nothing in the intellect which has not first been in the senses except the intellect itself*," and this has never been gainsaid by any school that could disprove it. The intuitionalist does not deny it: he merely claims that we are the recipients of another form of knowledge, the intuitional, which, instead of being derived from sense-experience, is projected into the intellectual consciousness from another source which we designate the Thinker. Thus, from the two forms of consciousness, come into the area of awareness truths that spring from entirely different sources. From the one source a steady stream of impressions flow constituting the substance of intellectual consciousness; from the other only a drop, every now and then, falls into the great inrushing mass so as to add a dim phosphorescence to an otherwise unilluminated pool. Obviously, when there is a lack of sensuous data from which a certain concept may be elaborated there can be no conception based upon them, and as the variety and quality of concepts are in exact proportion to the variety and quality of sense-experience there can be no demand for a particular species of notions such as might be elaborated out of the absent or non-existent perception. Hence, the power of conceiving springs forth from sense-experience. Sense-experience is essentially a mass of perceptions: these, creating a demand for additional adaptations, conspire, as if, to evoke the power or faculty to meet the demand, and consequently, an added conceptualization is made.

Progress in human thought is made in a manner similar to that which prevails in the development of other natural processes, such as, the power of speech in the child. In the development of this faculty there are certain definite stages which appear in due sequence. The child is not gifted with the power of speech at once. It comes, by gradual and sometimes painful growth, into a full use of this faculty. Now, much the same principle holds true in the evolution of the mind in the human species. It is an established biologic principle that the ontogenetic processes manifested in the individual are but a recapitulation of the phylogenetic processes which are observable in the progress of the entire species. The view becomes even more cogent when note is taken of the fact that the foetus, during embryogenesis, passes successively through stages of growth which have been shown to be analogous, if not identical, with those stages through which the human species has developed, namely, the mineral, vegetal and animal.

Wherefore it may be said that the fourth dimensional concept marks a distinct stage in psychogenesis or evolution of mind. It required, as will be shown in Chapter II, nearly two thousand years for it to germinate, take root and come to full fruition. For it was not until the early years of the nineteenth century that mathematicians, taking inspiration from RIEMANN (1826-1866) fully recognized the concept as a metaphysical possibility, or even the idea was conceived at all. Serious doubt is entertained as to the possibility of its conception by any human mind before this date, that is, the time when it was actually born. Prior to that time, mathematical thought was taking upon itself that shape and tendence which would eventually lead to the discovery of hyperspace; but it could not have reached the zenith of its upward strivings at one bound. That would have been unnatural.

Such is the constitution of the mind that although it is the quantity which bridges the chasm between the two stages of man's evolution when he merely thinks and when he really knows it is entirely under the domain of law and must observe the times and seasons, as it were, in the performance of its functions. The scope of psychogenesis is very broad, perhaps unlimited; but its various stages are very clearly defined notwithstanding the breadth of its scope of motility. And while the distance from *moneron* to man, or from feeling to thinking is vast, the gulf which separates man, the Thinker, from man, the knower, is vaster still. Who, therefore, can say what are the delights yet in store for the mind as it approaches, by slow paces, the goal whereat it will not need to struggle through the devious paths of perceiving, conceiving, analyzing, comparing, generalizing, inferring and judging; but will be able to know definitely, absolutely and instantaneously? That some such consummation as this shall crown the labors of mental evolution seems only natural and logical.

It may be thought by some that the character and content of revelational impressions constitute a variation from the requirements of the law above referred to, but a little thought will expose the fallacy of this view. The nature of a revealed message is such as to make it thoroughly amenable to the restrictions imposed by the evolutionary aspects of mind in general. That this is true becomes apparent upon an examination of the four cardinal characteristics of such impressions. First, we have to consider the indefinite character of an apocalyptic ideograph which is due to its symbolic nature. This is a feature which relieves the impression of any pragmatic value whatsoever, especially for the period embracing its promulgation. Then, such cryptic messages may or may not be understood by the recipient in which latter case it is nonpropagable. Second, the necessity of previous experience in the mind of the recipient in order that he may be able to interpret to his own mind the psychic impingement. The basis which such experience affords must necessarily be present in order that there may be an adequate medium of mental qualities and powers in which the ideogram may be preserved. A third characteristic is that revelations quite invariably presuppose a contemplative attitude of mind which, in the very nature of the case, superinduces a state of preparedness in the mind for the proper entertainment of the concept involved. This fact proves quite conclusively that revelational impressions are not exceptions to the general rule. Lastly, a dissatisfaction with the conditions with which the symbolism deals or to which it pertains is also a prerequisite. This condition is really that which calls forth the cryptic annunciation, and yet, preceding it is a long series of causes which have produced both the conditions and the revolt which the revelator feels at their presence. In view of the foregoing, it would appear that objections based upon the alleged nonconformity of the revealed or inspired cannot be entertained as it must be manifest that it, too, falls within the scope of the laws of mental growth.

Discoveries, whether of philosophical or mechanical nature, or whether of ethical or purely mathematical tendence, are never the results of a deliberate, methodical or purposive reflection. For instance, let us take LIE's "transformation groups," mathematic contrivances used in the solution of certain theorems. Now, it ought to be obvious that these mathetic machinations were not discovered by SOPHUS LIE as a consequence of any methodic or purposeful intention on his part. That is, he did not set out deliberately to discover "transformation groups." For back of the "groups" lay the entire range of analytic investigations; the mathematical thought of more than a thousand years furnished the substructure upon which Lie built the conception of his "groups." Similarly, it may be said with equal assurance that no matter how great the intensity of thought, nor how purposeful, nor of how long duration the series of concentrated abstractions which led up to the invention of the printing press, the linotype or multiplex printing press of our day could not have been produced abruptly, nor by use of the mental dynamics of the human mind of remoter days. Its production had to follow the path outlaid by the laws of psychogenesis and await the development of those powers which alone could give it birth. The whole question resolves itself, therefore, into the idea of the complete subserviency of the mind, in all matters of special moment, to the laws aforementioned. The supersession of the law of its own life by the mind is well-nigh unthinkable, if not quite so.

If we now view the history of the mind as manifested in the human species, three great epochs which divide the scope of mental evolution into more or less well-defined stages present themselves. These are: first, the *formative stage*; second, *the determinative stage*; third, the stage of *freedom*, or the *elaborative* stage.

In all of the early races of men, through every step which even preceded the *genus homo*, the generic mind was being formulated. It was being given shape, outline and direction. All of the first stage, the *formative*, was devoted to organization and direction. Those elementary sensations which constituted the basis of mind in the primitive man were accordingly strongly determinative of what the mind should be in these latter days. To this general result were contributed the effects of the activity of cells, nerves, bones, fibers, muscles and the blood.

The *formative* period naturally covered a very extensive area in the history of mind or psychogenetic development. It was followed closely, but almost insensibly, by the *determinative* period during which all the latent powers, capacities and faculties which were the direct products of the *formative* period were being utilized in meeting the demands of the law of necessity. The making of provisions against domestic want, against the attacks of external foes; the combating of diseases, physical inefficiency, the weather, wild beasts, the asperities of tribal enmities; as well as furthering the production of art, music, sculpture, the various branches of handiwork, literature, philosophies, religions and the effectuation of all those things which now appear as the result of the mental activity of the present-day man make up the essence and purpose of the determinative period.

Signs of the dawn of the *elaborative* stage, also called the stage of *freedom*, have been manifest now for upwards of three centuries and it is, therefore, in its beginnings. It is not fully upon us. Not yet can we fully realize what it may mean, nor can we unerringly forecast its ultimate outcome; but we feel that it is even now here in all the glories of its matutinal freshness. And the mind is beginning to be free from the grinding necessities of the constructive period having already freed itself from the restrictive handicaps of the primeval formulation period. Already the upgrowing rejuvenescences so common at the beginning of a new period are commencing to show themselves in every department of human activity in the almost universal desire for greater freedom. And this is particularly noticeable in the many political upheavals which, from time to time, are coming to the surface as well as in the countless other aspects of the wide-spread renaissance. Perhaps the time may come, never quite fully, when there will be no longer any necessity to provide against the external exigencies of life; perhaps, the time will never be when the mind shall no more be bound by the law of self-preservation, not even when it has attained unto the immortality of absolute knowledge; yet, it is intuitively felt that it must come to pass that the mind shall be vastly freer than it is to-day. And with this new freedom must come liberation from the necessities of the elementary problems of mere physical existence.

The inference is, therefore, drawn that the fourth dimensional concept, and all that it connotes of hyperspace or spaces of n-dimensionality are some of the evidences that this stage of freedom is dawning. And the mind, joyous at the prospect of unbounded liberty which these concepts offer, cannot restrain itself but has already begun to revel in the sunlit glories of a newer day. What the end shall be; what effect this new liberty will have on man's spiritual and economic life; and what it may mean in the upward strivings of the Thinker for that sublime perpetuity which is always the property of immediate knowledge no one can hope, at the present time, to fathom. It is, however, believed with KEYSER that "it is by the creation of hyperspaces that the rational spirit secures release from limitation"; for, as he says, "in them it lives ever joyously, sustained by an unfailing sense of infinite freedom."

The elevating influence of abstract thinking, such as excogitation upon problems dealing with entities inhabiting the domain of *mathesis* is, without doubt, incalculable in view of the fact that it is only through this kind of thought that the spirit is enabled to reach its highest possibilities. This is undoubtedly the philosophy of those religious and occult exercises known as "meditations," and this perhaps was the main idea in the mind of the Hebrew poet when he exclaimed: "Let the words of my mouth and the meditation of my heart be acceptable in thy sight, O Lord, my strength and my Redeemer." The principal, if not the only, value possessed by the "summitless hierarchies of hyperspaces" which the mathematician constructs in the world of pure thought is the enriching and ennobling influence which they exert upon the mind. But admittedly this unbounded domain of mathetic territory which he explores and which he finds "peopled with ideas, ensembles, propositions, relations and implications in endless variety and multiplicity" is quite real to him and subsists under a reign of law the penalties of which, while not as austere and unreasonable as some which we find in our tridimensional world, are nevertheless quite as palpable and as much to be feared. For the orthodoxy of mathematics is as cold and intolerant as ever the religious fanatic can be. But the reality and even the actuality which may be imputed to the domain of mathesis is of an entirely different quality from that which we experience in our world of triune dimensionality and it is a regrettable error of judgment to identify them. It ought, therefore, never be expected, nor is it logically reasonable to assume that the entities which inhabit the mathetic realm of the analyst should be submissive to the laws of sensible space; nor that the conditions which may be found therein can ever be made conformable to the conditions which exist in perceptual space.

It was PLATO's belief that ideas alone possessed reality and what we regard as actual and real is on account of its ephemerality and evanescence not real but illusionary. This view has been shared by a number of eminent thinkers who followed, with some ostentation, the lead established by PLATO. For a considerable period of time this school of thinkers had many adherents; but the principles at length fell into disrepute owing to the absurdities indulged in by some of the less careful followers. The realism, or for that matter, the actuality of ideas cannot be denied; yet it is a realism which is neither to be compared with the physical reality of sense-impressions nor its phenomena. The character and peculiarity of ideas are in a class apart from similar notions of perceptual space content. It is as if we were considering the potentialities of the spirit world and the entities therein in connection with incarnate entities which in the very nature of the case is not allowable. Furthermore, it is unreasonable to suppose that the conditions on a higher plane than the physical can be made responsible to a similar set of conditions on the physical plane.

There are certain astronomers who base their speculations as to the habitability of other planets upon the absurd hypothesis that the conditions of life upon all planets must be the same as those on the earth, forgetting that the extent of the universe and the scope of motility of life itself are of such a nature as to admit of endless variations and adaptations. There is a realism of ideas and a realism of perceptual space. Yet this is no reason why the two should be identified. On the other hand, owing to the diversity in the universe, every consideration would naturally lead to the assumption that they are dissimilar. To invest ideas, notions, implications and inferences with a reality need not logically or otherwise affect the reality of a stone, a fig, or even of a sense-impression.

To a being on the spirit levels our grossest realities must appear as non-existent. They are neither palpable nor contactable in any manner within the ordinary range of physical possibilities. For us his gravest experiences can have no reality whatsoever; for no matter how real an experience may be to him it is altogether beyond our powers of perception, and therefore, to us non-existent also. It should, however, be stated that the state of our knowledge about a given condition can in no way affect its existence. It merely establishes the fact that two or more realities may exist independent of one another and further that the gamut of realism in the universe is infinite and approaches a final state when its occlusion into absolute being follows as a logical sequence.

Recurring to the consideration of the reality of spirit-realms as compared with that of sensible space, it comes to view that our idealism, that is, the idealism which is a quality of conceptualization, may be regarded as identical with their realism, at least as being on the same plane as it. Stated differently, the things that are ideal to us and which constitute the data of our consciousness may be as real to them as the commonest object of sense-knowledge is to us. What, therefore, appears to us as the most ethereal and idealistic may have quite a realistic character for them.

Ultimately, however, and in the final deeps of analysis it will be found undoubtedly that both our realism and our idealism as well as similar qualities of the spirit world are in all essential considerations quite illusionary. All knowledge gained in a condition short of divinity itself is sadly relative. Even mathematical knowledge falls far short of the absolute, the fondest claims of the orthodox mathematician to the contrary notwithstanding. It has been said frequently that a mathematical fact is an absolute fact and that its verity, necessity and certainty cannot be questioned anywhere in the universe whether on Jupiter, Neptune, Fomalhaut, Canopus or Spica. But having so declared, the fact of the sheer relativity of our knowledge is not disturbed thereby nor controverted. Happily, neither distance nor a lack of distance can in any way affect the quality of human knowledge, mathematical knowledge not excepted. That can only be affected by conditions which cause it to approach perfection and nothing but evolution can do that.

In the light of results obtained in analytic investigations the question of the flexibility of mathematical applications becomes evident and one instead of being convinced of the vaunted invariability of the laws obtaining in the world of mathesis is, on the other hand, made aware of the remarkable and seemingly unrestrained facility with which these laws may be made to apply to any conditions or set of assumptions within the range of the mind's powers of conception. Mathematicians have deified the *definition* and endowed it with omnific powers imputing unto it all the attributes of divinity—immutability, invariance, and sempiternity. In this they have erred grievously although, perhaps, necessarily. Mathetic conclusions are entirely conditional and depend for their certainty upon the imputed certitude of other propositions which in turn are dependent, in ever increasing and endlessly complex relations, upon previously assumed postulates. These facts make it exceedingly difficult to understand the attitude of mind which has obscured the utter mutability and consequent ultimate unreliability of the fine-spun theories of analytic machinations.

The apriority of all mathematical knowledge is open to serious questioning. And although there is no hesitancy in admitting the basic agreement of the most primary facts of mathematical knowledge with the essential character of the intellect the existence of well-defined limits for such congruence cannot be gainsaid. The subjunctive quality of geometric and analytical propositions is made apparent by an examination of the possibilities falling within the scope of permissibility offered by mathetic license. For instance, privileged to proceed according to the analytic method it is allowable to reconstruct the sequence of values in our ordinary system of enumeration so as to admit of the specification of a new value for say, the entire series of odd numbers. This value might be assumed to be a plus-or-minus one, dependent upon its posture in the series. That is, all odd numbers in the series beginning with the digit 3, and continuing, 5, 7, 9, 11, 13, 15, 17, 19, ... n, could be assumed to have only a place value which might be regarded as a constant-variable. The series of even numbers, 2, 4, 6, 8, 10, 12, 14, 16, ... n, may be assumed to retain their present sequence values. Under this system the digit 1 would have an absolute value; all other odd numbers would have a constant-variable value; constant, because always no more nor less than 1 dependent upon their place in the operations and whether their values were to be applied by addition or subtraction to or from one of the values in the even number series; variable, because their values would be determinable by their application and algebraic use.

There would, of course, be utilitarian objection to a system of this kind; but under the conditions of a suppositionary hypothesis, it would be self-consistent throughout, and if given universal assent would suit our purposes equally as well as our present system. But the fact that this can be done under the mathematic method verily proves the violability of mathematical laws and completely negatives the assumption that the sum of any two digits, as say 2 plus 2 equals 4, is necessarily and unavoidably immutable. For it can be seen that the sum-value of all numbers may be made dependent upon the assumed value which may be assigned to them or to any collection thereof. Furthermore, it is a matter of historical knowledge that it was the custom of ancient races of men to account for values by an entirely different method from what we use to-day. The latter is a result of evolution and while experience teaches that it is by far the most convenient, it is nevertheless true that earlier men managed at least fairly well on a different basis. Then, too, the fact of the utility and universal applicability of our present system, based upon universal assent, does not obviate the conclusion that any other system, consistent in itself, might be made to serve our purposes as well.

It ought to be said, however, in justice to the rather utilitarian results obtained by La Grange, Helmholtz, Fechner, and others who strove to make use of their discoveries in analysis in solving mechanical, physiological and other problems of more or less pragmatic import that, in so far as this is true, mathematical knowledge must be recognized as being consistent with the necessities of *a priori* requirements. But even these results may not be regarded as transcending the scope of the most fundamental principles of sense-experience. It will be discovered finally, perhaps, that the energy spent in elaborating complicate series of analytic curiosities has been misappropriated. It will then be necessary to turn the attention definitely to the study of that which lies not at the terminus of the intellect's *modus vivendi,* but which is both the origin of the intellect and its eternal sustainer—the intuition, or life itself. This can result in nothing less than the complete spiritualization of man's mental outlook and the consequent inevitable recognition of the underlying and ever-sustaining *one-ness* of all vital manifestations.

One of the curiosities of the tendency in man's mind to specialize in analytics, whether in the field of pure mathematics or metaphysics, is the fact that it almost invariably leads to an attempt to account for cosmic origins on the basis of paralogic theories. This in times past has given rise to the theory of the purely mechanical origin of the universe as well as many other fantastic fallacies the chief error of which lay in the failure to distinguish between the realism of mental concepts and that of the sensible world. In spite of this, however, one is bound to appreciate the beneficial effects of analytic operations because they serve as invigorants to mental growth. It could not, therefore, be wished that there were no such thing as analytics; for the equilibria-restoring property of the mind may at all times be relied upon to minimize the danger of excesses in either direction. Just as the tide flowing in flows out again, thereby restoring the ocean's equilibrium, so the mind ascending in one generation beyond the safety mark has its equilibrium restored in the next by a relinquishment of the follies of the former.

The four-space is one of the curiosities of analytics; yet it need not be a menace to the sane contemplation of the variegated products of analysis. Safety here abides in the restraint which should characterize all discussion and application of the concept. If enthusiasts would be content not to transport the so-called fourth dimensional space out of the sphere of hyperspace and cease trying to speculate upon the results of its interposal into three space conditions, which is in every way a constructual impossibility, there could not be any possible objection to its due consideration. This would obviate the danger of calling into question either the sincerity or perspicacity of those whose enthusiasm tempts them to transgress the limits of propriety in their behavior towards the inquiry.

There is but one life, one mind, one extension, one quantity, one quality, one being, one state, one condition, one mood, one affection, one desire, one feeling, one consciousness. There is also but one number and that is unity. All so-called integers are but fractional parts of this kosmic unity. The idea represented by the word *two* really connotates two parts of unity and the same is true of a decillion, or any number of parts. These are merely the infinitesimals of unity and they grow less in size and consequence as the divisions increase in number. The analysis of unity into an infinity of parts is purely an *a posteriori* procedure. That it is an inherent mind-process is a fallacy. All our common quantities, as the mile, kilometer, yard, foot, inch, gallon, quart, are conventional and arbitrary and susceptible of wide variations. As the basis of all physical phenomena is unity; it is only in the ephemeral manifestations of sensuous objects that they appear as separate and distinct quantities.

We see on a tree many leaves, many apples or cherries; on a cob many grains of corn. We have learned to assign to each of these quantities in their summation a sequence value. But this is an empirical notion and cannot be said to inhere in the mind itself. Let us take, for instance, the mustard seed. If it were true that in one of these seeds there existed all the subsequent seeds which appear in the mustard plant as separate and identifiable quantities, and not in essence, then there would perhaps be warrant for the notion that diversity, as the calculable element, is an *a priori* conception. But, as this is not the case and since diversity is purely empirical and pertains only to the efflorescence of the one life it is manifestly absurd to take that view.

Under the most charitable allowances, therefore, there can be but two quantities—unity and diversity; yet not two, for these are one. Unity is the *one* quantity and diversity is the division of unity into a transfinity of parts. Unity is infinite, absolute and all-inclusive. Diversity is finite although it may be admitted to be transfinite, or greater than any assignable value. Unity alone is incomprehensible. In order to understand something of its nature we divide it into a diversity of parts; and because we fail to understand the transfinity of the multitude of parts we mistakenly call them infinite.

When analysis shall have proceeded far enough into the abysmal mysteries of diversity; when the mathematical mind shall have been overcome by the overwhelming perplexity of the maze of diverse parts, it shall then fall asleep and upon awaking shall find that wonderfully simple thing—*unity*. It is the one quantity that is endowed with a magnitude which is both inconceivable and irresolvable. The one ineluctable fact in the universe is the incomprehensibility and all-inclusivity of *one-ness*. It is incomprehensible, inconceivable and infinite at the present stage of mind development. But the goal of mind is to understand the essential character of unity, of life. Its evolution will then stop, for it will have reached the prize of divinity itself whereupon the intellect exalted by and united with the intuition shall also become one with the divine consciousness.

CHAPTER II

HISTORICAL SKETCH OF THE HYPERSPACE MOVEMENT

Egypt the Birthplace of Geometry—Precursors: NASIR-EDDIN, CHRISTOPH CLAVIUS, SACCHERI, LAMBERT, LA GRANGE, KANT—Influence of the *Mecanique Analytique*—The Parallel-Postulate the Root and Substance of the Non-Euclidean Geometry—The Three Great Periods: The Formative, Determinative and Elaborative—RIEMANN and the Properties of Analytic Spaces.

The evolution of the idea of a fourth dimension of space covers a long period of years. The earliest known record of the beginnings of the study of space is found in a hieratic papyrus which forms a part of the Rhind Collection in the British Museum and which has been deciphered by EISENLOHR. It is believed to be a copy of an older manuscript of date 3400 B. C., and is entitled "*Directions for Knowing All Dark Things*" The copy is said to have been made by AHMES, an Egyptian priest between 1700 and 1100 B. C. It begins by giving the dimensions of barns; then follows the consideration of various rectilineal figures, circles, pyramids, and the value of pi ([Greek: p]). Although many of the solutions given in the manuscript have been found to be incorrect in minor particulars, the fact remains that Egypt is really the birth-place of geometry. And this fact is buttressed by the knowledge that THALES, long before he founded the Ionian School which was the beginning of Greek influence in the study of mathematics, is found studying geometry and astronomy in Egypt.

The concept of hyperspace began to germinate in the latter part of the first century, B. C. For it was at this date that GEMINOS of RHODES (B. C. 70) began to think seriously of the mathematical labyrinth into which EUCLID's parallel-postulate most certainly would lead if an attempt at demonstrating its certitude were made. He recognized the difficulties which would engage the attention of those who might venture to delve into the mysterious possibilities of the problem. There is no doubt, too, but that EUCLID himself was aware, in some measure at least, of these difficulties; for his own attitude towards this postulate seems to have been one of noncommittance. It is, therefore, not strange that the astronomer, PTOLEMY (A. D. 87-165), should be found seeking to prove the postulate by a consideration of the possibilities of interstellar triangles. His researches, however, brought him no relief from the general dissatisfaction which he felt with respect to the validity of the problem itself.

For nearly one thousand years after the attempts at solving the postulate by GEMINOS and PTOLEMY, the field of mathematics lay undisturbed. For it was at this time that there arose a strange phenomenon, more commonly known as the "Dark Ages," which put an effectual check to further research or independent investigations. Mathematicians throughout this long lapse of time were content to accept EUCLID as the one incontrovertible, unimpeachable authority, and even such investigations as were made did not have a rebellious tendence, but were mainly endeavors to substantiate his claims.

Accordingly, it was not until about the first half of the thirteenth century that any real advance was made. At this time there appeared an Arab, NASIR-EDDIN (1201-1274) who attempted to make an improvement on the problem of parallelism. His work on EUCLID was printed in Rome in 1594 A. D., about three hundred and twenty years after his demise and was communicated in 1651 by JOHN WALLIS (1616-1703) to the mathematicians of Oxford University. Although his calculations and conclusions were respectfully received by the Oxford authorities no definite results were regarded as accomplished by what he had done. It is believed, however, that his work reopened speculation upon the problem and served as a basis, however slight, for the greater work that was to be done by those who followed him during the next succeeding eight hundred years.

About twenty years before the printing of the work of NASIR-EDDIN, CHRISTOPH CLAVIUS (1574) deduced the axiom of parallels from the assumption that a line whose points are all equidistant from a straight line is itself straight. In his consideration of the parallel-postulate he is said to have regarded it as EUCLID's XIIIth axiom. Later BOLYAI spoke of it as the XIth and later still, TODHUNTER treated it as the XIIth. Hence, there does not seem to have been any general unanimity of opinion as to the exact status of the parallel-postulate, and especially is this true in view of the uncertainty now known to have existed in EUCLID's mind concerning it.

GIROLAMO SACCHERI (1667-1733), a learned Jesuit, born at San Remo, came next upon the stage. And so important was his work that it will perpetuate the memory of his name in the history of mathematics. He was a teacher of grammar in the Jesuit *Collegio di Brera* where TOMMASO CEVA, a brother of GIOVANNI, the well-known mathematician, was teacher of mathematics. His association with the CEVA brothers was especially beneficial to him. He made use of CEVA's very ingenious methods in his first published book, 1693, entitled *Solutions of Six Geometrical Problems Proposed by Count Roger Ventimiglia.*

FIG. 1.

SACCHERI attacked the problem of parallels in quite a new way. Examining a quadrilateral, *ABCD,* in which the angles *A* and *B* are right angles and the sides *AC* and *BD* are equal, he determined to show that the angles *C* and *D* are equal. He also sought to prove that they are either right angles, obtuse or acute. He undertook to prove the falsity of the latter two propositions (that they are either obtuse or acute), leaving as the only possibility that they must be right angles. In doing so, he found that his assumptions led him into contradictions which he experienced difficulty in explaining.

His labors in connection with the solution of the problems proposed by COUNT VENTIMIGLIA, including his work on the question of parallels, led directly into the field of metageometrical researches, and perhaps to him as to no other who had preceded him, or at least to him in a larger degree, belongs the credit for a continued renewal of interest in that series of investigations which resulted in the formulation of the non-Euclidean geometry.

The last published work of SACCHERI was a recital of his endeavors at demonstrating the parallel-postulate. This received the "Imprimatur" of the Inquisition, July 13, 1733; the Provincial Company of Jesus took possession of the book for perusal on August 16, 1733; but unfortunately within two months after it had been reviewed by these authorities, SACCHERI passed away.

All efforts which had been made prior to the work of SACCHERI were based upon the assumption that there must be an equivalent postulate which, if it could be demonstrated, would lead to a direct, positive proof of EUCLID's proposition. Although these and all other attempts at reaching such a proof have signally failed and although it may correctly be said that the entire history of demonstrations aiming at the solution of the famous postulate has been one long series of utter failures, it can be asserted with equal certitude that it has proven to be one of the most fruitful problems in the history of mathematical thought. For out of these failures has been built a superstructure of analytical investigations which surpasses the most sanguine expectations of those who had labored and failed.

In 1766 JOHN LAMBERT (1728-1777) wrote a paper upon the *Theory of Parallels* dated Sept. 5, 1766, first published in 1786, from the papers left by F. BERNOULLI, which contained the following assertions:[2]

1. The parallel-axiom needs proof, since it does not hold for geometry on the surface of the sphere.

2. In order to make intuitive a geometry in which the triangle's sum is less than two right angles, we need an "imaginary" sphere (the pseudosphere).

3. In a space in which the triangle's sum is different from two right angles there is an absolute measure (a natural unit for length).

At this time IMMANUEL KANT (1724-1804), the noted German metaphysician, was in the midst of his philosophical labors. And it is believed that it was he who first suggested the idea of different *spaces*. Below is given a statement taken from his *Prolegomena*[3] which corroborates this view.

"That complete space (which is itself no longer the boundary of another space) has three dimensions, and that space in general cannot have more, is based on the proposition that not more than three lines can intersect at right angles in one point.... That we can require a line to be drawn to infinity, a series of changes to be continued (for example, *spaces* passed through by motion) in indefinitum, presupposes a representation of space and time which can only attach to intuition."

His differentiation between space in general and space which may be considered as the "boundary of another space" shows, in the light of the subsequent developments of the mathematical idea of space that he very fully appreciated the marvelous scope of analytic spaces. His conception of space, therefore, must have had a profound influence upon the mathematic thought of the day causing it to undergo a rapid reconstruction at the hands of geometers who came after him.

Under the masterly influence of LA GRANGE (1736-1813) the idea of different spaces began to take definite shape and direction; the geometry of hyperspace began to crystallize; and the field of mathesis prepared for the growth of a conception the comprehension of which was destined to be the profoundest undertaking ever attempted by the human mind. Unlike most great men whom the world learns tardily to admire, LA GRANGE lived to see his talents and genius fully recognized by his compeers; for he was the recipient of many honors both from his countrymen and his admirers in foreign lands. He spent twenty years in Prussia where he went upon the invitation of FREDERICK the Great who in the Royal summons referred to himself as the "greatest king in Europe" and to LA GRANGE as the "greatest mathematician" in Europe. In Prussia the *Mecanique Analytique* and a long series of memoirs which were published in the Berlin and Turin Transactions were produced. LA GRANGE did not exhibit any marked taste for mathematics until he was 17 years of age. Soon thereafter he came into possession of a memoir by HALLEY quite by accident and this so aroused his latent genius that within one year after he had reviewed HALLEY's memoir he became an accomplished mathematician.

He created the calculus of variations, solved most of the problems proposed by Fermat, adding a number of theorems of his own contrivance; raised the theory of differential equations to the position of a science rather than a series of ingenious methods for the solution of special problems and furnished a solution for the famous isoperimetrical problem which had baffled the skill of the foremost mathematicians for nearly half a century. All these stupendous tasks he performed by the time he reached the age of nineteen.

The *Mecanique Analytique* is his greatest and most comprehensive work. In this he established the law of virtual work from which, by the aid of his calculus of variations, he deduced the whole of mechanics, including both solids and liquids. It was his object in the *Analytique* to show that the whole subject of mechanics is implicitly embraced in a single principle, and to lay down certain formulae from which any particular result can be obtained. He frequently made the assertion that he had, in the *Mecanique Analytique*, transformed mechanics which he persistently defined as a "geometry of four dimensions"[4] into a branch of analytics and had shown the so-called mechanical principles to be the simple results of the calculus. Hence, there can be no doubt but that LA GRANGE not only completed the foundation, but provided most of the material in his analyses and other "abstract results of great generality" which he obtained in his numerous calculations, for the superstructure subsequently known as the geometry of hyperspace, and in which the fourth dimensional concept occupies a very fundamental place.

It is as if for nearly seventeen hundred years workmen, such as GEMINOS, of RHODES, PTOLEMY, SACCHERI, NASIR-EDDIN, LAMBERT, CLAVIUS, and hundred of others who struggled with the problem of parallels, had made more or less sporadic attempts at the excavation of the land whereon a marvelously intricate building was to be constructed. There is no historical evidence to show that any of them ever dreamed that the results of their labors would be utilized in the manner in which they have been used. Then came KANT with the wonderfully penetrating searchlight of his masterful intellect who from the elevation which he occupied saw that the site had great possibilities, but he had not the mathematical talent to undertake the work of actual, methodical construction. Indeed his task was of a different sort. However, he succeeded in opening the way for LA GRANGE and others who followed him. LA GRANGE immediately seized upon the idea which for more than a thousand years had been impinging upon the minds of mathematicians vainly seeking lodgment and began the elaboration of a plan in accordance with which minds better skilled in the pragmatic application of abstract principles than his could complete the work begun. Unfortunately, on account of his intense devotion and loyalty to the study of pure mathematics, and when he had reached the summit of his greatness where he stood "without a rival as the foremost living mathematician," his health became seriously affected, causing him to suffer constant attacks of profound melancholia from which he died on April 10, 1813.

We come now to one of the most remarkable periods in the history of mental development. During the six hundred years between the birth of NASIR-EDDIN and the death of LA GRANGE the entire world of mathesis was being reconstituted. Since there had been gradually going on an internal process which, when completed, forever would liberate the mind from the narrow confines of consciousness limited to the three-space, it is not surprising that we should find, in the mathematical thought of the time, an absolutely epoch-making departure. The innumerable attempts at the solution of the parallel-postulate, all failures in the sense that they did not prove, have intensified greatly the esteem in which the never-dying elements of EUCLID are held to-day. And despite the fact that there may come a time when his axioms and conclusions may be found to be incongruent with the facts of sensuous reality; and though all of his fundamental conceptions of space in general, his theorems, propositions and postulates may have to give way before the searching glare of a deeper knowledge because of some revealed fault, the perfection of his work in the realm of pure mathematics will remain forever a master piece demanding the undiminished admiration of mankind.

The parallel-postulate, as stated by EUCLID in his *Elements of Geometry*, reads as follows:

> "If a straight line meet two straight lines so as to make the two interior angles on the same side of it taken together less than two right angles, these straight lines being continually produced, shall at length meet upon that side on which are the angles which are less than two right angles."

On this postulate hang all the "law and the prophets" of the non-Euclidean Geometry. In it are the virtual elements of three possible geometries.

Furthermore, it is both the warp and the woof of the loom of present-day metageometrical researches. It is the golden egg laid by the god SEB at the beginning of a new life cycle in psychogenesis. Its progeny are numerous—hyperspaces, sects, straights, digons, equidistantials, polars, planars, coplanars, invariants, quaternions, complex variables, groups and many others. A wonderfully interesting breed, full of meaning and pregnant with the power of final emancipations for the human intellect!

When the conclusions which were systematically formulated as a result of the investigations along the lines of hypotheses which controverted the parallel-postulate were examined it was found that they fell into three main divisions, namely: the synthetic or hyperbolic; the analytic or RIEMANNIAN and the elliptic or CAYLEY-KLEIN. These divisions or groups are based upon the three possibilities which inhere in the conception taken of the sum of the angles referred to in the above postulate as to whether it is equal to, greater or less than two right angles.

The assumption that the angular sum is congruent to a straight angle is called the Euclidean or parabolic hypothesis and is to be distinguished from the synthetic or hyperbolic hypothesis established by GAUSS, LOBACHEVSKI and BOLYAI and which assumes that the angular sum is less than a straight angle. The elliptic or CAYLEY-KLEIN hypothesis assumes that the angular sum is greater than a straight angle. LOBACHEVSKI, however, not satisfied with the statement of the parallel-postulate as given by EUCLID and which had caused the age-long controversy, substituted for it the following:

"All straight lines which, in a plane, radiate from a given point, can, with respect to any other straight line, in the same plane, be divided into two classes—the intersecting and the non-intersecting. The boundary line of the one and the other class is called parallel to the given line."

This is but another way of saying about the same thing that EUCLID had declared before, and yet, curiously enough it afforded just the liberty that LOBACHEVSKI needed to enable him to elaborate his theory.

For the purposes of this sketch the field of the development of non-Euclidean geometry is divided into three periods to be known as: (1) the *formative* period in which mathematical thought was being formulated for the new departure; (2) the *determinative* period during which the mathematical ideas were given direction, purpose and a general tendence; (3) the *elaborative* period during which the results of the former periods were elaborated into definite kinds of geometries and attempts made at popularizing the hypotheses.

The Formative Period

CHARLES FREDERICH GAUSS (1777-1855) by some has been regarded as the most influential mathematician that figured in the formulation of the non-Euclidean geometry; but closer examination into his efforts at investigating the properties of a triangle shows that while his researches led to the establishment of the theorem that a regular polygon of seventeen sides (or of any number which is prime, and also one more than a power of two) can be inscribed, under the Euclidean restrictions as to means, in a circle, and also that the common spherical angle on the surface of a sphere is closely connected with the constitution of the area inclosed thereby, he cannot justly be designated as the leader of those who formulated the synthetic school. And this, too, for the simple reason that, as he himself admits in one of his letters to TAURINUS, he had not "published anything on the subject." In this same letter he informs TAURINUS that he had pondered the subject for more than thirty years and expressed the belief that there could not be any one who had "concerned himself more exhaustively with this second part (that the sum of the angles of a triangle cannot be more than 180 degrees)" than he had.

Writing from Göttingen to TAURINUS, November 8, 1824, and commenting upon the geometric value of the sum of the angles of a triangle, he says:

> "Your presentation of the demonstration that the sum of the angles of a plane triangle cannot be greater than 180 degrees does, indeed, leave something to be desired in point of geometrical precision. But this could be supplied, and there is no doubt that the impossibility in question admits of the most rigorous demonstration. But the case is quite different with the second part, namely, that the sum of the angles cannot be smaller than 180 degrees; this is the real difficulty, the rock upon which all endeavors are wrecked.... The assumption that the sum

of the three angles is smaller than 180 degrees leads to a new geometry entirely different from our Euclidean—a geometry which is throughout consistent with itself, and which I have elaborated in a manner entirely satisfactory to myself, so that I can solve every problem in it with the exception of the determining of a constant which is not *a priori* obtainable."

It appears from this correspondence that GAUSS had in the privacy of his own study elaborated a complete non-Euclidean geometry, and had so thoroughly familiarized himself with its characteristics and possibilities that the solution of every problem embraced within it was very clear to him except that of the determination of a constant. He concluded the above letter by saying:

"All my endeavors to discover contradiction or inconsistencies in this non-Euclidean geometry have been in vain, and the only thing in it that conflicts with our reason is the fact that if it were true there would necessarily exist in space a linear magnitude quite determinate in itself; yet unknown to us."

Judging from the correspondence between GAUSS and GERLING (1788-1857), BESSEL (1784-1846), SCHUMACHER and TAURINUS, the nephew of SCHWEIKART, and that between SCHWEIKART and GERLING, there had grown up a general dissatisfaction in the minds of mathematicians of this period with Euclidean geometry and especially the parallel-postulate and its connotations. BESSEL expresses this general discontent in one of his letters to GAUSS, dated February 10, 1829, in which he says:

"Through that which LAMBERT said and what SCHWEIKART disclosed orally, it has become clear to me that our geometry is incomplete, and should receive a correction, which is hypothetical, and if the sum of the angles of the plane triangle is equal to 180 degrees, vanishes."

The opinion of leading mathematicians at this time seems to have been crystallizing very rapidly. Unconsciously the men of this formative period were adducing evidence which would give form and tendence to the developments in the field of mathesis at a later date. They appear to have been reaching out for that which, ignis fatuus-like, was always within easy reach, but not quite apprehensible.

A bolder student than GAUSS was FERDINAND CARL SCHWEIKART (1780-1857) who also has been credited with the founding of the non-Euclidean geometry. In fact, if judged by the same standards as GAUSS, he would be called the "father of the geometry of hyperspace"; for he really published the first treatise on the subject. This was in the nature of an inclosure which he inserted between the leaves of a book he loaned to GERLING. He also asked that it be shown to GAUSS that he might give his judgment as to its merits.

SCHWEIKART's treatise, dated Marburg, December, 1818, is here quoted in full:

"There is a two-fold geometry—a geometry in the narrower sense, the Euclidean, and an astral science of magnitude.

"The triangles of the latter have the peculiarity that the sum of the three angles is not equal to two right angles.

"This presumed, it can be most rigorously proven: (*a*) That the sum of the three angles in the triangle is less than two right angles.

"(*b*) That this sum becomes ever smaller, the more content the angle incloses. (*c*) That the altitude of an isosceles right-angled triangle indeed ever increases, the more one lengthens the side; that it, however, cannot surpass a certain line which I call the constant."

Squares have consequently the following form:

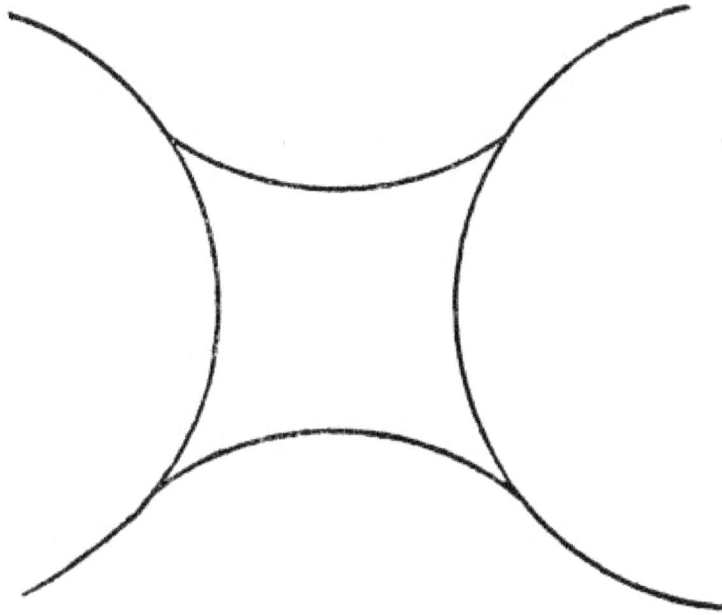

FIG. 2.

"If this constant were for us the radius of the earth (so that every line drawn in the universe, from one fixed star to another, distant 90° from the first, would be a tangent to the surface of the earth) it would be infinitely great in comparison with the spaces which occur in daily life."

The above, being the first published, not printed, treatise on the new geometry occupies a unique place in the history of higher mathematics. It gave additional strength to the formative tendencies which characterized this period and marked SCHWEIKART as a constructive and original thinker.

The nascent aspects of this stage received a fruitful contribution when NICOLAI LOBACHEVSKI (1793-1847) created his *Imaginary Geometry* and JANOS BOLYAI (1802-1860) published as an appendix to his father's *Tentamen,* his *Science Absolute of Space.* LOBACHEVSKI and BOLYAI have been called the "Creators of the Non-Euclidean Geometry." And this appellation seems richly to be deserved by these pioneers. Their work gave just the impetus most needed to fix the status of the new line of researches which led to such remarkable discoveries in the more recent years. The *Imaginary Geometry* and the *Science Absolute of Space* were translated by the French mathematician, J. HOÜEL in 1868 and by him elevated out of their forty-five years of obscurity and non-effectiveness to a position where they became available for the mathematical public. To Bolyai and Lobachevski, consequently, belong the honor of starting the movement which resulted in the development of metageometry and hence that which has proved to be the gateway of a new mathematical freedom.

Gauss, Schweikart, Lobachevski, Wolfgang and Janos Bolyai were the principal figures of the formative period and the value of their work with respect to the formulation of principles upon which was constructed the Temple of Metageometry cannot be overestimated.

The Determinative Period

This period is characterized chiefly by its close relationship to the theory of surfaces. Riemann's Habilitation Lecture on *The Hypotheses Which Constitute the Bases of Geometry* marks the beginning of this epoch. In this dissertation, Riemann not only promulgated the system upon which Gauss had spent more than thirty years of his life in elaborating, for he was a disciple of Gauss; but he disclosed his own views with respect to space which he regarded as a particular case of manifold. His work contains two fundamental concepts, namely, the *manifold* and the *measure of curvature* of a continuous manifold, possessed of what he called *flatness* in the smallest parts. The conception of the measure of curvature is extended by Riemann from surfaces to spaces and a new kind of space, finite, but unbounded, is shown to be possible. He showed that the dimensions of any space are determined by the number of measurements necessary to establish the position of a point in that space. Conceiving, therefore, that space is a manifold of finite, but unbounded, extension, he established the fact that the passage from one element of a manifold to another may be either discrete or continuous and that the manifold is discrete or continuous according to the manner of passage. Where the manifold is regarded as discrete two portions of it can be compared, as to magnitude, by counting; where continuous, by measurement. If the whole manifold be caused to pass over into another manifold each of its elements passing through a one-dimensional manifold, a two-dimensional manifold is thus generated. In this way, a manifold of n-dimensions can be generated. On the other hand, a manifold of n-dimensions can be analyzed into one of one dimension and one of $(n-1)$ dimensions.

To RIEMANN, then, is due the credit for first promulgating the idea that space being a special case of manifold is generable, and therefore, *finite*. He laid the foundation for the establishment of a special kind of geometry known as the "elliptic." Space, as viewed by him, possessed the following properties, viz.: generability, divisibility, measurability, ponderability, finity and flexity.

These are the six pillars upon which rests the structure of hyperspace analyses.[5]

Generability is that property of geometric space by virtue of which it may be generated, or constructed, by the movement of a line, plane, surface or solid in a direction without itself. *Divisibility* is that property of geometric space by virtue of which it may be segmented or divided into separate parts and superposed, or inserted, upon or between each other. *Measurability* is that property by virtue of which geometric space is determined to be a manifold of either a positive or negative curvature, also by which its extent may be measured. *Ponderability* is that property of geometric space by virtue of which it may be regarded as a quantity which can be manipulated, assorted, shelved or otherwise disposed of. *Finity* is that property by virtue of which geometric space is limited to the scope of the individual consciousness of a unodim, a duodim or a tridim and by virtue of which it is finite in extent. *Flexity* is that property by virtue of which geometric space is regarded as possessing curvature, and in consequence of which progress through it is made in a curved, rather than a geodetic line, also by virtue of which it may be flexed without disruption or dilatation.

RIEMANN who thus prepared the way for entrance into a veritable labyrinth of hyperspaces is, therefore, correctly styled "The father of metageometry," and the fourth dimension is his eldest born. He died while but forty years of age and never lived long enough fully to elaborate his theory with respect to its application to the measure of curvature of space. This was left for his very energetic disciple, EUGENIO BELTRAMI (1835-1900) who was born nine years after RIEMANN and lived thirty-four years longer than he. His labors mark the characteristic standpoint of the determinative period. BELTRAMI'S mathematical investigations were devoted mainly to the non-Euclidean geometry. These led him to the rather remarkable conclusion that the propositions embodied therein relate to figures lying upon surfaces of constant negative curvature.

BELTRAMI sought to show that such surfaces partake of the nature of the pseudosphere, and in doing so, made use of the following illustration:

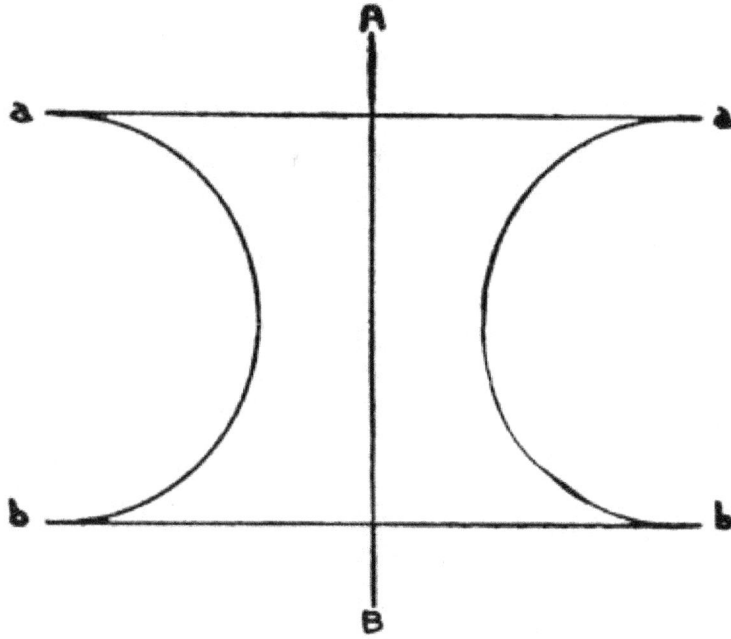

FIG. 3.

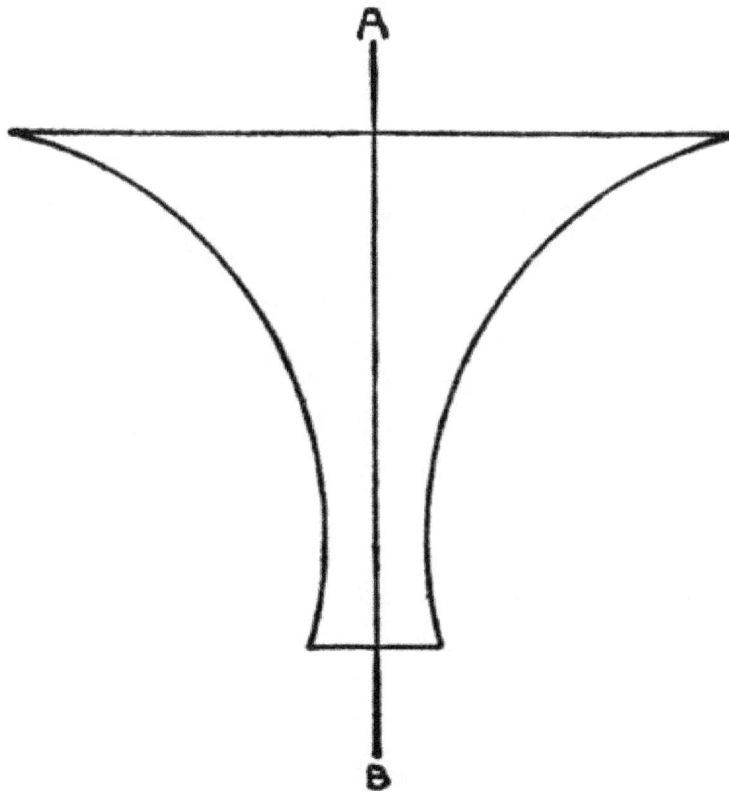

FIG. 4.

If the plane figure *aabb* is made to revolve upon its axis of symmetry *AB* the two arcs, *ab* and *ab* will describe a pseudospherical concave-convex surface like that of a solid anchor ring. Above and below, toward *aa* and *bb*, the surface will turn outward with ever-increasing flexure till it becomes perpendicular to the axis and ends at the edge with one curvature infinite. Or, the half of a pseudospherical surface may be rolled up into the shape of a champagne glass, as in Fig. 4. In this way, the two straightest lines of the pseudospherical surface may be indefinitely produced, giving a kind of space (pseudospherical) in which the axiom of parallels does not hold true.

The determinative period marks the most important stage in the development of non-Euclidean geometry and certainly the most significant in the evolution of the idea of hyperspaces and multiple dimensionality. RIEMANN and BELTRAMI are chief among those whose labors characterize the scope of this period. Their work gave direction and general outline for later developments and all subsequent researches along these lines have been conducted in strict conformity with the principles laid down by these pioneer constructionists. They laid out the field and designated its confines beyond which no adventurer has since dared to pass.

The great importance of the work of RIEMANN at this time may be seen further in the fact that it not only marked the beginning of a new epoch in geometry; but his pronouncement of the hypothesis that space is unbounded, though finite, is really the first time in the history of human thought that expression was ever given to the idea that space may yet be only of limited extent. Before that time the minds of all men seemed to have been unanimous in the consideration of space as an illimitable and infinite quantity.

The Elaborative Period

The elaborative stage includes the work of all those who, working upon the bases laid down by LOBACHEVSKI, BOLYAI, SCHWEIKART and RIEMANN, have sought to amplify the conclusions reached by them. Among those whose investigations have greatly multiplied the applications of hyperspace conceptions are HOÜEL (1866) and FLYE ST. MARIE (1871) of France; HELMHOLTZ (1868), FRISCHAUF (1872), KLEIN (1849), and BALTZER (1877) of Germany; BELTRAMI (1872) of Italy; DE TILLY (1879) of Belgium; CLIFFORD and CAYLEY (1821) of England; NEWCOMB (1835) and HALSTEAD of America.

These have been most active in popularizing the subject of non-Euclidean geometry and incidentally the idea of the fourth dimension. The great mass of non-professional mathematical readers, therefore, owe these men an immeasurable debt of gratitude for the work that they have done in the matter of rendering the conceptions which constitute the fabric of metageometry understandable and thinkable. A glance at the bibliography appended at the end of this volume will give some idea of the enormous amount of labor that has been expended in an effort to translate the most abstract mathematical principles into a language that could easily be comprehended by the average intelligent person.

The characteristic standpoint of this period is the popular comprehension of the hyperspace concept and the consequent mental liberation which follows. For there is no doubt but that unheard of possibilities of thought have been revealed by investigations into the nature of space. An entirely new world has been opened to view and only a beginning has been made at the exploration of its extent and resources.

One of the notable incidents of the early years of this period is the position taken by FELIX KLEIN who stands in about the same relation to CAYLEY as BELTRAMI does to RIEMANN, in that he assumed the task of completing the work of his predecessor. KLEIN held that there are only two kinds of RIEMANNIAN *space*—the elliptical and the spherical. Or in other words, that there are only two possible kinds of space in which the propositions announced by RIEMANN could apply. SOPHUS LIE, called the "great comparative anatomist of geometric theories," carried his classifications to a final conclusion in connection with spaces of all kinds and decided that there are possible only four kinds of three dimensional spaces.

But whether men with peering, microscopic, histological vision shall establish the existence of one or many spaces, and regardless of the mathematic rigor with which they shall demonstrate the self-consistency of the doctrines which they hold, the fact remains that the hypotheses thus maintained, while they may be regarded as true descriptions of the spaces concerned, are, nevertheless, incompatible. All of them cannot be valid. It will perhaps be found that none of them are valid, especially objectively so. The only true view, therefore, of these systems of hyperspaces is that which assigns them to their rightful place in the infinitely vast world of pure mathesis where their validity may go unchallenged and their existence unquestioned; for in that domain of unconfined mentation, in that realm of divine intuitability, the marvelous wonderland of ideas and notions, one is not only disinclined to doubt their logical actuality, but is quite willing to accede their claims.

CHAPTER III

ESSENTIALS OF THE NON-EUCLIDEAN GEOMETRY

The Non-Euclidean Geometry Concerned with Conceptual Space Entirely—Outcome of Failures at Solving the Parallel-Postulate—The Basis of the Non-Euclidean Geometry—Space Curvature and Manifoldness—Some Elements of the Non-Euclidean Geometry—Certainty, Necessity and Universality as Bulwarks of Geometry—Some Consequences of Efforts at Solving the Parallel-Postulate—The Final Issue of the Non-Euclidean Geometry—Extended Consciousness.

The term "non-Euclidean" is used to designate any system of geometry which is not strictly Euclidean in content.

It is interesting to note how the term came to be used. It appears to have been employed first by GAUSS. He did not strike upon it suddenly, however, as in the correspondence between him and WACHTER in 1816 he used the designation "anti-Euclidean" and then, later, following SCHWEIKART, he adopted the latter's terminology and called it "Astral Geometry." This he found in SCHWEIKART's first published *treatise* known by that name and which made its appearance at Marburg in December, 1818. Finally, in his correspondence with TAURINUS in 1824, GAUSS first used the expression "non-Euclidean" to designate the system which he had elaborated and continued to use it in his correspondence with SCHUMACHER in 1831.

"Non-Legendrean," "semi-Euclidean" and "non-Archimedean" are titles used by M. DEHN to denote all kinds of geometries which represented variations from the hypotheses laid down by LEGENDRE, EUCLID and ARCHIMEDES.

The semi-Euclidean is a system of geometry in which the sum of the angles of a triangle is said to be equal to two right angles, but in which one may draw an infinity of parallels to a straight line through a given point. The non-Euclidean geometry embraces all the results obtained as a consequence of efforts made at finding a satisfactory proof of the parallel-postulate and is, therefore, based upon a conception of space which is at variance with that held by EUCLID. According to the Ionian school space is an infinite continuum possessing uniformity throughout its entire extent. The non-Euclideans maintain that space is not an infinite extension; but a finite though unbounded manifold capable of being generated by the movement of a point, line or plane in a direction without itself. It is also held that space is curved and exists in the shape of a sphere or pseudosphere and is consequently elliptical.

The inapplicability of EUCLID's parallel-postulate to lines drawn upon the surface of a sphere suggested the possibility of a space in which the postulate could apply to all possible surfaces or that space itself may be spherical in which case the postulate would be invalidated altogether. Hence, it is quite natural that mathematicians finding themselves unable to prove the postulate with due mathetic precision should turn their attention to the conceptually possible. In this virtual abandonment of the perceptual for the conceptual lies the fundamental difference between the Euclidean and the non-Euclidean geometries. It may be said to the credit of the Euclideans that they have sought to make their geometric conceptions conform as closely as possible to the actual nature of things in the sensuous world while at the same time they must have perceived that at best their spatial notions were only approximations to the sensuous actuality of objects in space.

On the other hand, non-Euclideans make no pretense at discovering any congruency between their notions and things as they actually are. The attitude of the metageometricians in this respect is very aptly described by CASSIUS JACKSON KEYSER who says:

> "He constructs in thought a summitless hierarchy of hyperspaces, an endless series of orderly worlds, worlds that are possible and logically actual, and he is content not to know if any of them be otherwise actual or actualized."[6]

The non-Euclidean is, therefore, not concerned about the applicability of ensembles, notions and propositions to real, perceptual space conditions. It is sufficient for him to know that his creations are thinkable. As soon as he can resolve the nebulosity of his consciousness into the conceptual "star-forms" of definite ideas and notions, he sits down to the feast which he finds provided by superfoetated hypotheses fabricated in the deeps of mind and logical actualities imperturbed and unmindful of the weal of perceptual space in its homogeneity of form and dimensionality.

Fundamentally, the non-Euclidean geometry is constructed upon the basis of conceptual space almost entirely. Knowledge of its content is accordingly derived from a superperceptual representation of relations and interrelations subsisting between and among notions, ideas, propositions and magnitudes arising out of a conceptual consideration thereof. In other words, representations of the non-Euclidean magnitudes, cannot be said to be strictly perceptual in the same sense that three-space magnitudes are perceived; for three-space magnitudes are really sense objects while hyperspace magnitudes are not sense objects. They are far removed from the sensuous world and in order to conceive them one must raise his consciousness from the sensuous plane to the conceptual plane and become aware of a class of perceptions which are not perceptions in the strict sense of the word, but superperceptions; because they are representations of concepts rather than precepts.

Notions of perceptual space are constituted of the triple presentations arising out of the visual, tactual and motor sensations which are fused together in their final delivery to the consciousness. The synthesis of these three sense-deliveries is accomplished by equilibrating their respective differences and by correcting the perceptions of one sense by those of another in such a way as to obtain a completely reliable perception of the object. This is the manner in which the characteristics of Euclidean space are established.

The characteristics of non-Euclidean space are not arrived at exactly in this way. Being beyond the scope of the visual, tactile and motor sense apprehensions, it cannot be said to represent judgments derived from any consideration or elaboration of the deliveries presented through these media. Yet, the substance of metageometry, or the science of the measurement of hyperspaces, may not be regarded as an *a priori* substructure upon which the system is founded. That is, the conceptual space of non-Euclidean geometry is not presented to the consciousness as an *a priori* notion. On the other hand, the *a posterioristic* quality of metageometric spaces marks the entire scope of motility of the notions appertaining thereto.

The notions, therefore, of conceptual space are derivable only from the perception of concepts, or, otherwise consist of judgments concerning interconceptual relations. The process of apperception involved in the recognition of relations which may be methodically determined is much removed from the primary procedure of perceiving sense-impressions and fusing them into final deliveries to the consciousness for conceptualization or the elaboration into concepts or general notions. It is a procedure which is in every way superconceptual and extra-sensuous. The metageometrician or analyst in no way relies upon sense-deliveries for the data of his constructions; for, if he did, he should, then, be reduced to the necessity of confining his conclusions to the sphere of motility imposed by the sensible world with the result that we should be able to verify empirically all his postulations. But, contrarily, he goes to the extra-sensuous, and there in the realm of pure conceptuality, he finds the requisite freedom for his theories; thus, environed by a sort of intellectual anarchism, he pursues analytical pleasures quite unrestrainedly. The difference between the two mental processes—that which leads from the sensible world to conception and that which veers into the fields beyond—is so great that it is hardly permissible to view the results arrived at in the outcome of the separate processes as being identical.

To illustrate this difference, let us draw an analogy. The miner digs the iron ore out of the ground. The iron is separated from the extraneous material and delivered to the furnaces where the metal is melted and turned out as pig iron. It is further treated, and steel, of various grades, cast iron and other kinds of iron are produced. The treatment of the iron ore up to this stage is similar to the treatment of sense-impressions by the Thinker. Steel, cast iron, et cetera, are similar to mental concepts. Later, the steel and other products are converted into instruments and numerous articles. This represents the superperceptual process. Trafficking in iron ore products, such as instruments of precision, watch springs, and the like, represents a stage still farther removed from the primary treatment of the ore and is similar to that to which concepts are treated when the metageometrician manipulates them in the construction of conceptual space-forms. Perception is the dealing with raw iron ore while conception is analogous to the production of the finished product.

Superperception would be analogous to the trafficking in the finished product as such and without any reference to the source or the preceding processes. Thus the notions and judgments of the non-Euclidean geometry are arrived at as a result of a triple process of perception, conception and superperception the latter being merely superconceived as formal space-notions. But it is obvious that the more complex the processes by which judgments purporting to relate to perceptual things are derived the more likely are those judgments to be at variance with the nature of the things themselves.

In view of the foregoing, the dangers resulting from identifying the products of the two processes are very obvious indeed. But the difference between the two procedures is the difference between Euclidean and non-Euclidean geometries or the difference between perceptual space notions and conceptual space notions. Hence, it is not understood just how or why it has occurred to anyone that the two notions could be made congruent. Magnitudes in perceptual, sensible space are things apart from those that may be said to exist in mathematical space or that space whose qualities and properties have no existence outside of the mind which has conceived them. It is believed to be quite impossible to approach the study of metageometrical propositions with a clear, open mind without previously understanding the fundamental distinctions which exist between them.

It follows, therefore, as a logical conclusion that geometric space of whatsoever nature is a purely formal construction of the intellect, and for this reason is completely under the sovereignty of the intellect however whimsical its demands may be. Being thus the creature of the intellect, its possibilities are limited only by the limitations of the intellect itself. Perceptual space, being neither the creature of the intellect nor necessarily an *a priori* notion resident in the mental substructure, but existing entirely independent of the intellect or its apprehension thereof, cannot be expected to conform to the purely formal restrictions imposed by the mind except in so far as those restrictions may be determined by the nature of perceptual space. And for that matter, it should not be forgotten that, as yet, we have no means of determining whether or not the testimony of the intellect is thoroughly credible simply because there is no other standard by which we may prove its testimony. It is possible to justify the deliveries of the eye by the sense of touch, or vice versa. It is also possible to prove all our sense-deliveries by one or the other of the senses. But we have no such good fortune with the deliveries of the intellect. We have simply to accept its testimony as final; because we cannot do any better. But if it were possible to correct the testimony of the intellect by some other faculty or power which by nature might be more accurate than the intellect it should be found that the intellect itself is sadly limited.

The possible curvature of space is a notion which also characterizes the content of the non-Euclidean geometry. It is upon this notion that the question of the finity and unboundedness of space, in the mathematical sense, rests. In the curved space, the straightest line is a curved line which returns upon itself. Progression eastward brings one to the west; progression northward brings one to the south, et cetera. On this view space is finite, but may not be regarded as possessing boundaries.

Space-curvature, reinforced by the idea that space is also a manifold is the enabling clause of metageometry and without them the analyst dares not proceed. Here again, we are led to the confession that however fantastic these two notions may seem and evidently are, there is nevertheless to be recognized in them a "dim glimpse" of a veritable reality—a slight foreshadowing of the revelation of some great kosmic mystery.

The manifoldness of space is the fiat of analysis. It is the inevitable outcome of the analyst's method of procedure. His education, training and view of things in general inhibit his arriving at any other result and he may be pardoned with good grace for his manufacture of the space-manifold. For by it perhaps a better appreciation of that wonderful extension of consciousness in the nature of which is involved the explanation of the perplexing problems which the manifold and other metageometrical expedients faintly adumbrate may be gained.

It is pertinent, in the light of the above, to examine into some of the relative merits of the three formal bulwarks of geometrical knowledge. These are *certainty*, *necessity* and *universality*.

Geometric certainty is derived solely from the nature of the premises upon which it is based. If the premises be contradictory, it is, of course, defective. But if the premises are non-contradictory or self-evident, then the certainty of geometric notions and conclusions is valid. Another consideration of prime importance in this connection is the *definition*. From it all premises proceed. Hence, the definition is even more important than the premise; for it is the persisting determinant of all geometric conclusions while the premise is dependent upon the limitations of the definition. The determinative character of the definition has led to its apotheosis; but this, admittedly, has been necessary in order to give stability and permanency to the conclusions which followed. But in spite of this it would appear that the certainty of geometric conclusions is not a quality to be reckoned as absolute or final.

With the same certainty that it can be said the sum of the angles of the triangle is equal to two right angles it may be asserted that that sum is also greater or less than two right angles. Certainty which is based upon the inherent congruity of definitions, premises and propositions is an entirely different matter from that certainty which arises out of the real, abiding validity of a scheme of thought. But this difference is not lessened by the fact that the latter is dependent, in a measure, upon the correct systematization of our spatial experiences by means of methodical processes. Euclidean geometry, accordingly, is not so certain in its applications as it is utilitarian; but non-Euclidean geometry is even less certain than the former and consequently more lacking in its utilitarian possibilities.

The necessity of geometrical determinations is merely the necessity which inheres in logical inferences or deductions. These may or may not be valid. Inasmuch as the necessariness of deductions is primarily based upon the conditional certainty of premises and definitions it appears that this quality is in no way peculiar to geometry whether Euclidean or non-Euclidean. In like manner, the universality of geometric judgments may not properly be regarded as a peculiarity of geometry; but is explicable upon the basis of the formal character of the assumptions which underlie it. The chief value, then, of non-Euclidean geometry seems to abide in the fact that it clarifies our understanding as to the complex processes by which it is possible to organize and systematize our spatial experiences for assimilation and use in other branches of knowledge.

With the above statement of the case of the non-Euclidean geometry it is now thought permissible to state briefly some of the elements thereof.[7]

Below will be found some of the elements obtained as a consequence of efforts made both at proving and disproving the parallel-postulate of Euclid:

"If two points determine a line it is called a straight."

"If two straights make with a transversal equal alternate angles they have a common perpendicular."

"A piece of a straight is called a sect."

"If two equal coplanar sects are erected perpendicular to a straight, if they do not meet, then the sect joining their extremities makes equal angles with them and is bisected by a perpendicular erected midway between their feet."

"The sum of the angles of a rectilineal triangle is a straight angle, in the hypothesis of the right (angle); is greater than a straight angle in the hypothesis of the obtuse (angle); is less than a straight angle in the hypothesis of the acute (angle)."

"The hypothesis of right is Euclidean; the hypothesis of the acute is BOLYAI-LOBACHEVSKIAN; the hypothesis of obtuse is RIEMANNIAN."

"If one straight is parallel to a second the second is parallel to the first."

"Parallels continually approach each other."

"The perpendiculars erected at the middle point of the sides of a triangle are all parallel, if two are parallel."

"If the foot of a perpendicular slides on a straight its extremity describes a curve called an equidistant curve, or an equidistantial."

"An equidistantial will slide on its trace."

"In the hypothesis of the obtuse a straight is of finite size and returns into itself."

"Two straights always intersect."

"Two straights perpendicular to a third straight intersect at a point half a straight from the third either way."

"A pole is half a straight from its polar."

"A polar is the locus of coplanar points half a straight from its pole. Therefore, if the pole of one straight lies on another straight the pole of this second straight is on the first straight."

"The cross of two straights is the pole of the join of their poles."

"Any two straights inclose a plane figure, a digon."

"Two digons are congruent if their angles are equal."

"The equidistantial is a circle with center at the poles of its basal straight."

A typical postulate based upon the BOLYAI hypothesis of the acute angle is the following:

"From any point P drop PC, a perpendicular to any given straight line AB. If D move off indefinitely on the ray CB, the sect will approach as limit PF copunctal with AB at infinity.

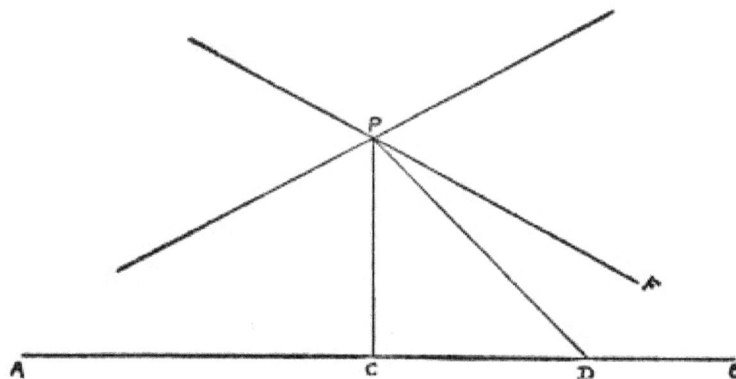

FIG. 5.

PD is said to be at P the parallel to AB toward B. PF makes with PC an angle CPF which is called the angle of parallelism for the perpendicular

PC. It is less than a right angle by an amount which is the limit of the deficiency of the triangle *PCD*. On the other side of *PC*, an equal angle of parallelism gives the parallel *P* to *BA* towards *AM*.[8] Thus at any point there are two parallels to a straight. A straight has, therefore, two separate points at infinity."

"Straights through *P* which make with *PC* an angle greater than the angle of parallelism and less than its supplement do not meet the straight *AB* at all not even at infinity."

The parallel-postulate is stated in the non-Euclidean geometry as follows:

"If a straight line meeting two straight lines make those angles which are inward and upon the same side of it less than two right angles the two straight lines being produced indefinitely will meet each other on this side where the angles are less than two right angles."

It is stated by MANNING[9] in the following language:

"If two lines are cut by a third and the sum of the interior angles on the same side of the cutting line is less than two right angles the line will meet on that side when sufficiently produced."

It is rather significant that in this postulate which is really a definition of space should be found grounds for such diverse interpretations as to its nature. Of course, the moment the mind seeks to understand the infinite by interpreting it in the unmodified terms of the apparently unchangeable finite it entangles itself into insurmountable difficulties. As a drowning man grasps after straws so the mind, immersed in endless abysses of infinity, fails to conduct itself in a seemly manner; but gasps, struggles and flounders and is happy if it can, in the depths of its perplexity, discover a way of logical escape. The pure mathematician has a hankering after the logically consistent in all his pursuits; to him it is the "Holy Grail" of his highest aspirations. He seeks it as the devotee seeks immortality. It is to him a philosopher's stone, the elixir of perpetual youth, the eternal criterion of all knowledge.

Failures to demonstrate the celebrated postulate of EUCLID led, as a matter of course, to the substitution of various other postulates more or less equivalent to it in that each of them may be deduced from the other without the aid of any new hypothesis.

Among those who sought proof by a restatement of the problem are the following:

1. PTOLEMY: The internal angles which two parallels make with a transversal on the same side are supplementary.

2. CLAVIUS: Two parallel straight lines are equidistant.

3. PROCLUS: If a straight line intersects one of two parallels it also intersects the other.

4. WALLIS: A triangle being given another triangle can be constructed similar to the given one and of any size whatever.

5. BOLYAI (W.): Through three points not lying on a straight line a sphere can always be drawn.

6. LORENZ: Through a point between the lines bounding an angle a straight line can always be drawn which will intersect these two lines.

7. SACCHERI: The sum of the angles of a triangle is equal to two right angles.

There were, of course, many other statements and substitutions used by mathematicians in their endeavors satisfactorily to establish the truth of the parallel-postulate. That their labors should have terminated, first, by doubting it, then by denying, and finally, by building up a system of geometries which altogether ignores the postulate is just what might naturally be expected of these men who have given to the world the non-Euclidean geometry. In doing what they did many, if not all of them, were not aware in any measure of the proportions of the imposing superstructure that would be built upon their apparent failures. All of them undoubtedly must have sensed the vague adumbrations forecast by the unfolding mysteries which they sought to lay bare; all of them must have felt as they executed the early tasks of those crepuscular days of pure mathematics that the way which they were traveling would lead to the inner shrine of a higher knowledge and a wider freedom; they may have been led by divine intuition to strike out on this new path and yet they could not have known how fully their dreams would be realized by the mathematicians of the twentieth century. If so, they were truly gods and mathesis is their kingdom.

The analyst proceeds upon a basis entirely at variance with that which guides the ordinary investigator in the formulation of his conclusions. The empirical scientist in arriving at his theories or hypotheses is governed at all times by the degree of conformity which his postulates exhibit to the actual phenomena of nature. He endeavors to ascertain just how far or in what degree his hypothesis is congruent with things found in nature. If the dissidence is found to predominate he abandons his theory and makes another statement and again sets out to determine the degree of conformity. If he then finds that the natural phenomena agree with his theory he accepts it as for the time being finally settling the question. In all things he is limited by the answer which nature gives to his queries. Not so with the exponent of pure mathematics. For him the truth of hypotheses and postulates is not dependent upon the fact that physical nature contains phenomena which answer to them. The sole determining factor for him is whether or not he is able to state with *rational consistency* the assumed first principles and then logically develop their consequences. If he can do this, that is, if he can state his hypotheses with consistency and develop their consequences into a logical system of thought, he is quite satisfied and well pleased with his performances. But the fact that this is true is of vital significance for all who seek clearly to understand the essential character of hyperspatiality.

It appears, therefore, that the science of consequences is the radical essence of pure geometry. The metageometrician enjoys unlimited freedom in the choice of his postulates and suffers curtailment only when it comes to the question of consistency. He is at liberty to formulate as many systems of geometry as the barriers of consistency will permit and these are practically innumerable. So long then as the laws of compatibility remain inviolate his multiplication of postulate-systems may proceed indefinitely. Is it strange then that under conditions where an investigator has such unbridled liberty he should be found indulging in mathetic excesses?

KANT held that the axioms of geometry are synthetic judgments *a priori*; but it appears that in the strictest sense this is not the case. It depends upon the type of mind which is taken as a standard of reference. If it be the uncultivated mind, it is certain that to it the relations expressed by an axiom would never appear spontaneously. If on the other hand, the standard be that of a cultivated mind it is also equally certain that to it these relations would be discovered only after methodical operations. All judgments arrived at as a result of logical processes should, it seems, be regarded as judgments *a posteriori*, i.e., the results of empirical operations. Confessedly, the facts adduced in course of experimentation serve as guides in choosing among all of the many possible logical conventions; but our choice remains untrammeled except by the compulsion arising out of a fear of inconsistency. The real criterion then of all geometries is neither truth, conformability nor necessity, but consistency and convenience.

The difficulty with the non-Euclideans resolved itself into the question as to whether it is more consistent, as well as convenient, to establish a proof of the postulate by taking advantage of the support to be found in other postulates or whether, by seeking a demonstration based upon the deliveries of sense-experience as to the nature of space and its properties, a still more consistent conclusion might be reached. They had further perplexity, however, when it came to a decision as to whether the organic world is produced and maintained in Euclidean space or in a purely conceptual space which alone can be apprehended by the mind's powers of representation. Unwilling to admit the existence of the world in Euclidean space, they turned their attention to the examination of the properties of another kind of space so-called which unlike the space of the Ionian school could be made to answer not only all the purposes of plane and solid figures, but of spherics as well. And so, the manifold space was invented by RIEMANN and later underwent some remarkable improvements at the hands of his disciple, BELTRAMI. But it may be said here, parenthetically, that the truth of the whole matter is that our world is neither in Euclidean nor non-Euclidean space, both of which, in the last analysis, are conceptual abstractions. Although it may not be denied that the Euclidean space is the more compatible.

The problem of devising a space, if only a very limited portion, in which could be demonstrated the assumed alternative hypothesis and its consequences logically developed, occasioned no inconsiderable concern for the non-Euclidean investigators; but neither LOBACHEVSKI, BOLYAI nor RIEMANN were to be baffled by the difficulties which they met. These only cited them to more laborious toil. Having succeeded in mentally constructing the particular kind of space which was adaptable to their rigorous mathetic requirements it immediately occurred to them that all the qualities of the limited space thus devised might logically be amplified and extended to the entire world of space and that what is true of figures constructed in the segmented portion of space which they used for experimental purposes is also true of figures drawn anywhere in the universe of this space as all lines drawn in the finite, bounded portion could be extended indefinitely and all magnitudes similarly treated. From these results, it was but a single step to the conclusion which followed—that either an entirely new world of space had been discovered or that our notion of the space in which the organic world was produced is wholly wrong and needs revision. But notwithstanding the insurmountable obstacles which stood in the way of the investigators who made the attempt to discover the homology which might exist between the characteristics of the newly fabricated space and the phenomenal world, investigations were carried forward with almost amazing recklessness and loyalty to the mathetic spirit until it was discovered that all efforts to trace out any definite lines of correspondence were futile. Then the policy of ignoring the question of conformability was adopted and has since been pursued with unchangeable regularity by the analytical investigator.

Among the results obtained by the non-Euclideans in their profound researches into the nature of hyperspace are these: 1. It was found that the angular sum of a triangle, being ordinarily assumed to be a variable quantity, is either less or greater than two right angles so that a strictly Euclidean rectangle could not be constructed. 2. The angle sums of two triangles of equal area are equal. 3. No two triangles not equal can have the same angles so that similar triangles are impossible unless they are of the same size. 4. If two equal perpendiculars are erected to the same line, their distance apart increases with their length. 5. A line every point of which is equally distant from a given straight line is a curved line. 6. Any two lines which do not meet, even at infinity, have one common perpendicular which measures their minimum distance. 7. *Lines which meet at infinity are parallel.* But it is apparent that these results have not followed upon any mathematical consequence of other supporting postulates or axioms such as would place them on a coördinate basis with those used as a support for the parallel-postulate; for they are based upon the envisagement of an entirely new principle of space-perception and belong to a wholly different set of space qualities.

The final issue then of the non-Euclidean geometry is neither in the utility of its processes and conclusions nor in the increscent inclination towards a new outlook upon the world of mathesis; but resides solely in the possibilities yet to be developed in that vast domain of analytical thought which it has discovered and opened to view. To say that it sheds any light upon the nature of the universe is perhaps to take the radical view; yet it cannot be doubted that the researches incident to the formulation of the non-Euclidean geometry have greatly extended the scope of consciousness. Whether the extension is valid and normal or simply a hypertrophic excrescence of mental feverishness; whether by virtue of it we shall more closely approach an understanding of the true nature of the mind of the Infinite, or shall all fall into insanity, are certainly debatable questions. It nevertheless appears evident that humanity has gained something of real, abiding permanence by this new departure. If that something be merely an extended consciousness or an awakening to the fact that there are stages of awareness beyond the strictly sensuous, and every observable evidence points to this, then there has only begun the process by which the faculty of conscious functioning in this new world shall become the normal possession of the human species. But this new world cannot be said to be of mathematical import; for it is doubtful if mathematical laws such as have been devised up to the present time, would obtain therein. So that if anything, it must be psychological and vital.

On this view the worlds of hyperspace inlaid with analytic manifoldnesses and constant curvatures are but the primal excitants which will finally awaken in the mind the faculty of awareness in the new domain of psychological content. Then will come the blooming of the diurnal flower of the mind's immortality and the outputting of the organ of consciousness wherewith the infinite stretches of hyperspaces, the low-lying valleys of reals and imaginaries and the uplifting hills of finites and infinites shall be divested of their mysteries and stand out in their unitariness no longer draped in the veil of the inscrutable and the incomprehensible.

The fourth dimension, regarded by some as a new scope of motion for objects in space, by others as a new and strange direction of spatial extent and by others still as the doorway of the temple of exegesis wherein an explanation may be found for the entire congeries of mysteries and supermysteries which now perplex the human mind, may also be said to be the key to the non-Euclidean geometry. But it really complicates the situation; for one has to be capable of prolonged abstract thought even to envisage is as a conceptual possibility. POINCARÉ[10] says: "Any one who should dedicate his life to it could, perhaps, eventually imagine the fourth dimension," implying thereby that a lifetime of prolonged abstract thought is necessary to bring the mind to that point of ecstasy where it could even so much as imagine this additional dimension. Nevertheless by it (the fourth dimension) was the non-Euclidean geometry made and without it was not any of the hyperspaces made that were made. It is the view which geometers have taken of space in general that has made the fourth dimension possible, and not only the fourth, but dimensions of all degrees. The basis of the non-Euclidean geometry may be found then in the notion of space which has been predominant in the minds of the investigators.

Finally, it should be pointed out that the non-Euclidean geometry, though a consistent system of postulates, has been constructed upon a misconception based upon the identification of real, perceptual space with systems of space-measurements. Hyperspaces which are not spaces at all should not be confounded with *real space*. But they constitute the substance of non-Euclidean geometry; they are its blood and sinews. Their study is interesting, because of the possibilities of speculation which it offers. No mind that has thought deeply upon the intricacies of the fourth dimension, or hyperspace, remains the same after the process. It is bound to experience a certain sense of humility, and yet some pride born of a knowledge that it has been in the presence of a great mystery and has delved into the fearful deeps of kosmic mind. To the mind that has thus been anointed by the sacred chrism of the inner mysteries of creative mentality there always come that stillness and calm such as characterize the aftermath of reflection upon the incomprehensible and the transfinite.

CHAPTER IV

DIMENSIONALITY

Arbitrary Character of Dimensionality—Various Definitions of Dimension—Real Space and Geometric Space Differentiated—The Finity of Space—Difference Between the Purely Formal and the Actual—Space as Dynamic Appearance—The *A Priori* and the *A Posteriori* as Defined by PAUL CARUS.

In previous chapters we have traced the growth and development of the non-Euclidean geometry showing that the so-called fourth dimension is an aspect thereof. It is now deemed fitting that we should enter into a more detailed study of the question of dimensionality with a view to examining some of the difficulties which encompass it.

The question of dimension is as old as geometry itself. Without it geometric conclusions are void and meaningless. Yet the conception of dimensionality itself is purely conventional. In its application to space there is involved a great deal of confusion because of the inferential character of its definition. For instance, commonly we measure a body in space and arbitrarily assign three elements to determine its position. The simplest standard for this purpose is the cube having three of its edges terminating at one of its corners.

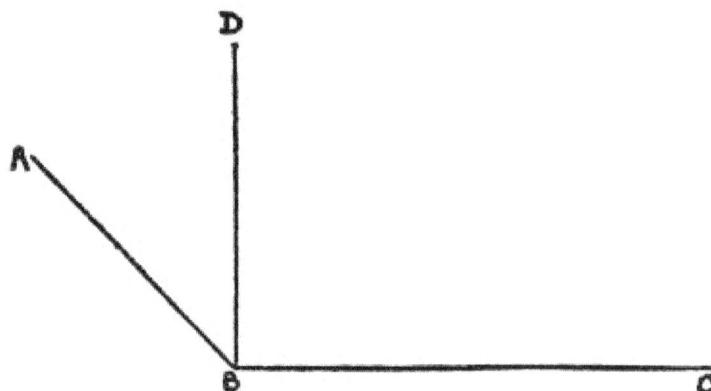

FIG. 6.

Thus because it is found that the entire volume of a cube is actually comprehended within the directions indicated by the lines *ab, bc* and *db* it is determined that the three coördinates of the point *b* are necessary and sufficient to establish the dimensions of the cube and consequently of the space in which it rests. The conception may be stated in this way: If a collection of elements, say points or lines, be of such a nature or order that it is sufficient to know a certain definite number of facts about it in order to be able to distinguish every one of the elements from all the others, then the assemblage or collection of elements is said to be of the same number of dimensions as there are elements necessary to its determination. In the above figure there are three elements, namely, the lines *ab, bc,* and *db,* which are necessary and sufficient for the determination of the position of the point *b.* In this way geometers have determined that our space is tridimensional; but it is obvious that this conclusion is based not upon any examination of space itself but upon the measurement of bodies in space. Upon this view it is seen that conclusions based upon such a procedure render our notion of the extension of bodies in space identical with the notion of spatial extensity. In other words, we take bodies in space and by examining their characteristics and properties arrive at an alleged apodeictic judgment of space. It is by means of this conventional norm of geometric knowledge that various other spaces, notably the one-, two-, four-and *n*-space, have been devised. It would appear that if some more absolute standard of measurement or definition of space were adopted the confusion which now clings to the conception of dimension could be obviated. For if it be true that three and only three elements are necessary to determine a point-position in our space and that in this determination we also find the number of dimensions of space, then it may also be true that *n*-coördinates would just as truly determine the dimensionality of an *n*-space, which is granted. But then the *n*-space would be just as legitimate as the three-space; for it is determined by exactly the same standards. It is both quantitatively and qualitatively the same. If, however, on account of the exigencies that might arise, we are forced to seek solace in the notion of an *n*-space whither shall we turn for it? It cannot be found; for it is imperceptible, uninhabitable, non-existent, and therefore, absolutely and purely an abstraction. Consequently, there must be something radically wrong with the definition of space or with its determinants.

The purely arbitrary character of dimensionality is very aptly described by Cassius Jackson Keyser, who says:

"... The dimensionality of a given space is not unique, but depends upon the choice of the geometric entity for primary or generating element. A space being given, its dimensionality is not therewith determined, but depends upon the will of the investigator who by a proper choice of generating element endows the space with any dimensionality he pleases. That fact is of cardinal significance for science and philosophy."[11]

It is a fact of "cardinal significance" for science; because it emphasizes the necessity for some more rational procedure than that of the geometrician in arriving at an absolutely unique method of determining the dimension and essential nature of real space. Its significance for philosophy lies in the need of a logical, rigidly exclusive and absolutely peculiar standard of space definition. The definition of perceptual space should be such as rigorously inhibits its inclusion as a particular in any general class. The necessity for this is warranted by its universality and uniqueness.

The lines of demarkation between what is recognized as perceptual space and what has been called geometric or conceptual space should be very sharply drawn. So that when reference is made to either there will be no doubt as to which is meant. And then, too, conceptual space is no space at all, properly speaking. It is merely a system of space-measurement. And as such has no logical right to be put in the same category as perceptual space.

Real space is unique. Geometric space belongs to a class whose members are capable of indefinite multiplication. It is certainly most illogical to identify them. Perceptual space, figuratively speaking, is a quantity; analytic space is the foot-rule, the yard-stick, the kilometer, by which it is measured and apportioned. It is logically impossible to predicate the same conclusion for both of them. That is, to do so causes a profound fracture of the fundamental norms of logic. Such conclusions being thus illegitimate it is rather surprising that an error of this nature should have been made. It is perhaps accountable for on the grounds of the geometer's complete *insouciance* as to how his postulates shall stand in their relation to things in the phenomenal world.

It is agreed that as convenient as is Euclid's system of space-measurement it is not by any means congruent with the extension of real space objects. It does, however, approximate congruity with these objects as nearly as possible. How then could it be expected that a system of space-measurement so far removed from this primary congruence as the non-Euclidean system is should exhibit more obvious signs of correspondence? But the advocates of the *n*-dimensionality of space have illatively asserted the identity of space and its dimensions. Accordingly, there is not recognized any distinction between their conception of space itself and its qualitative peculiarities. They use the terms interchangeably. So that dimension means space and *vice versa*. In this lack of discrimination may be found the source of much of the confusion which attaches to the conception of space.

If it were arguable that the relation between space and its dimensions is the same as that between matter and its properties then the restriction of this relation to three and only three directions of extent would be disallowed; for the reason that if, as is commonly done, dimension be made to mean direction of extent, there would be an unlimited number of directions of extent and they would all be perceptible. But this is really another fundamental fault. Non-Euclideans have stretched the meaning of the term dimension so that it not only includes the idea of direction but an entirely new class of qualities—the fourth dimension. And despite this reformation of the original conception, they demand that it shall be called space.

We have just shown that the generic concept of dimensionality is that three and only three coördinates are necessary and sufficient for its determination. Granting that this is true, are we not compelled consequently to see that we have, by adding a fourth or *n*-dimensions, involved ourselves into a more complex situation than before? For by postulating a fourth dimension either we have created a new world whose dimensions are four in number or we have explicitly admitted that the three dimensions have a fourth. Aside from the logical difficulties which beset these conclusions there is also set up a condition which is at variance with the most elementary requirements of common sense.

Thus far mathematical thought has not served to clarify our notions of space nor to shed any new light upon the vital processes which are alleged to have their explanation in the new discovery. Simply stated, metageometricians have brought us to the place where we must either recognize that the fourth dimension is another sphere lying dangerously near the earth in which space extends in four primary directions and in which four coördinates are necessary for its determination or we are driven to the other horn of the dilemma where we are brought face to face with the conclusion that the three perceptual space dimensions have in common a hitherto unknown property or extension in virtue of which it may be viewed as having an unlimited number of dimensions. To accept the latter view is equivalent to saying that, in the above figure, the three lines *ab*, *bc* and *db* have formed a triple *entente* by which they have mutually and severally acquired a new domain, hyperspace, and in which, because of the vast resources of the region, they are able to perform wondrous things.

Let us examine briefly the various current definitions of dimension. It is assumed by not a few that dimension is the same as *direction*. But can we grant this wholly to be true? If so, then a mere child may see that there are and must necessarily be as many dimensions as there are directions. Primarily, there are six directions of space and an unlimited number of subsidiary directions. On this view it is not necessary to invent a new domain of space if the object be merely to discover and utilize a greater number of dimensions than has heretofore been allowed. For the identification of the term dimension with direction already makes available an almost infinite number of dimensions. But this view is objected to by the advocates, for it is contrary to the hypothesis of *n*-dimensionality.

Dimension also means *extent*. This is partially true. It cannot be wholly true. For, if it were, then, space would have only one dimension which is also not allowable under the hypothesis. Then the definition leaves out of account the idea that space is at the same time a direction or collection of directions. The term extension is generic and when applied to space means extension in all possible directions and not in any one direction. So that it is not permissible to say that space extends in this direction or that because it extends in all directions simultaneously and equally.

Geometers claim that space is a system of coördinates necessary for the establishment of a point-position in it. This view, however, identifies space with a system of space-measurement and is therefore faulty. According to this view there may be as many spaces as there are systems of space-measurement and the latter may be limitless. But if the totality of spaces are to be viewed as one space then we shall have one space with an indefinite number of dimensions; also an indefinite number of space measurements which would be confusing. Much, if not all, of such a system's utility and convenience would be unavailable or useless. That, too, would be in violation of the avowed purpose of these investigations which is to enhance the utility and convenience of mathematic operations.

Now it is evident that space is neither direction, extension, a system of space-measurement nor a system of manifolds whose dimensions are generable. And this is so for the same reason that a piece of cloth is not the elements of measurement—inches, feet, yards—by which it is apportioned. And because we find that the fabric of space lends itself accommodatingly to our conventional norms of measurement is not sufficient reason for identifying it with these norms. Here we have the source of all error in mathematical conclusions about the nature of space; because all such conclusions are based not upon the intrinsic nature of space, but upon artificial forms which we choose to impose upon it for our own convenience. But it should be remembered that the irregularities which we note are not in space itself but inhere in the forms which we use. For these purposes space is extremely elastic and accommodates itself to the shape and scope of any construction we may decide to try upon it. In this respect it is like water which has no regard for the shape, size or kind of vessel into which it may be posited. There is one thing certain that judging from the above considerations there has been not yet any absolute, all-satisfying definition devised for space by mathematicians.

The best definitions hitherto constructed are purely artificial and arbitrary determinations. It is rather anomalous that there should be so little unanimity about what is the most fundamental consideration of mathematical conclusions which are supposed to be so certain, so necessary and universal as to be incontrovertible. Confessedly, it is a condition which raises again the question as to just what are the limits of mathematical certainty and necessity and just how far we shall depend upon the validity of mathematics to determine for us absolutely certain conclusions about the nature of space. In view of the uncertainty noted, are we justified in following too closely the mathematic lead even in matters of logic, to say nothing of our conception of space? It seems that we shall have necessarily, on account of the recognized limitations of mathematics in this matter, to turn to some more tenable source for the norms of our knowledge concerning space. For in the light of the rather indefensible position which metageometricians have involved themselves there appears to be no hope in this direction.

It is undoubtedly safer not to rely altogether upon the purely abstract, even in the world of mathesis, for any absolute criterion of knowledge. It is perhaps well that we should expunge the word *absolute* from our vocabularies. It is really a misnomer and has no meaning in the lexicon of nature. There is in reality no *absolute* in the sense of final absolution from all conditions or restrictions.

In the ultimate analysis there is unquestionably no hue, tone, quality, condition nor any imaginable posture of life, being or manifestation that is absolved from every other one of its class or from the totality. All these are relational and interdependent. There is no room for the absolute. In fact, it is a quality which cannot in any way be ascribed to any aspect of kosmic manifestation. It has existence only in the mind and has been devised for the purpose of marking the limits of its scope. All being is relative; all life is relative and is destined to change its qualities as it evolves. All knowledge is also relative and what is true of one state may not be true of another; what is true of one life may not be true of another life; the limitations of one degree of knowledge may not have any bearings upon another degree. The norms of one will not satisfy the conditions of another stage of manifestation. It is always within limits that the criterion of knowledge will be found to satisfy a given set of conditions. Hence within certain limits mathematical conclusions will maintain their validity. Error is committed by pushing the validity of these limits to a position without the sphere of limitations. This seems to be the crux of the whole matter. Mathematicians, notably non-Euclideans, have sought to extend the comparatively small sphere of limits of congruence between mathematic and perceptual space to such an extent as to cause it to encroach upon forbidden territory. In doing this they have erred grievously, causing serious offense to the more sensitive spirit of the high-caste mathematicians among whom are none more truly conservative than PAUL CARUS,[12] who says:

"Metageometricians are a hot-headed race and display sometimes all the characteristics of sectarian fanatics. To them it is quite clear there may be two straight lines through one and the same point which do not coincide and yet are both parallel to a third line."

To the student who has carefully followed the development of the non-Euclidean geometry and the notion of hyperspace the above characterization is none too severe nor ill-deserved. Nothing could more vividly yet correctly portray the impious tactics of the metageometrician and establish his perceptual obliquity more surely than the mere fact, mentioned by CARUS, that he can with evident lack of mental perturbation proclaim that two straight lines, noncoincident with each other, may pass through a point and yet be parallel to a third line. But this is a mere trifle, a bagatelle, to the many other infractions of which he is guilty. The wonder is that he is able to secure such obsequious acceptance of his offerings as many of the most serious minded mathematicians are inclined to give. Is it to be wondered at that, despite the profuse protestations of the advocates, many who take up the study of the question of hyperspace should experience a deep revulsion from the posture assumed by metageometricians with respect to these queries?

Linked with the idea of dimensionality is the notion that space is infinite. This is a conception which has its roots imbedded in the depths of antiquity. Primitive man, looking up into the heavens at what appeared to him as a never ending extension, was awed by its vastness; but the minds of the most learned of the present-day men are not free from this innate dread of infinity. It permeates the thought life of all alike and none seems to be able to rise above it. Mathematicians, philosophers, scientists all share in the general belief that space is without limit, unending in extent and eternally existent. RIEMANN, whose thought life found its most convenient mode of expression by means of pure mathematics, was the first in the history of human thought to surmise that space is not infinite but limited even though unbounded. But his conception has been much vitiated on account of its entanglement with an *idealized* construction by which space is regarded as a thing to be manipulated and generated by act of thought. Were it not for this his conception would indeed mark the beginning of a new era in psychogenesis. As it is, when all the nonsensical effusions have been cleared away from our space conceptions and men come really to understand something of the essential nature of space this new era will find its true beginnings in the mind of RIEMANN. Although it must be said, as is the case with all progressive movements, the later development of a rationale for this conclusion will vary greatly from his original conception. For he had in mind a space that is generable and therefore a logical construction while ultimately the mind will swing back to a consideration of real space.

Already men are beginning to see a new light. Already they are beginning to take a new view of space in general. The departure is especially noticeable in the attitude assumed by HIRAM M. STANLEY.[13] He says:

"If we seek the most satisfactory understanding of space we shall look neither to mathematics nor Psychology but to Physics. The trend of Physics, say with such a representative as OSTWALD, is to make things the expression of force; the constitution and appearance of things are determined by dynamism; and we may best interpret space as a mode of this dynamic appearance."

Space, as a mode of dynamic appearance is a slight improvement upon the old idea of a pure vacuity; for in the light of what we now know about space content much of the dignity of that view has been lost. Men now know that space is not an empty void. They know that the atmosphere fills a great deal of space. They also have extended their conception in this direction to include the ether and occultism goes further and postulates four kinds of ether—the chemical, life, light and psychographic ethers. But it does not stop here. It postulates a series of grades of finer matter than the physical which fills space and permeates its entire extent even to identification with its essential nature.

STANLEY continues:

"Everything does not, as commonly conceived, fall into some pre-existent space convenient for it; but everything makes its own spaciousness by its own defensive and offensive force, and the totality of all appearance is space in general."

According to STANLEY, not only do physical, perceptual objects, by their "offensive and defensive force" make their own space but the appearance of that in which no physical object is makes room for itself by its own dynamic force. In other words, that which we call "pure extensity" is by virtue of its dynamism the cause of its own existence.

At first hand there appears to be little worthy of serious consideration in this view of STANLEY; yet, if carried to its logical conclusion, the merit of the hypothesis becomes apparent. Accordingly, interstellar distances which are commonly said to be even without air or life of any kind are really an appearance possessed of a dynamism peculiar to itself. And this very force-appearance, constituting space, is that which makes it perceivable. For instance, let us say the space that exists between the earth and the moon, is not really empty nor does it have an existence prior to itself, but is a mode of dynamic appearance which is the cause of its own existence. Its dynamic character makes it to appear perceptible to our senses. Logically, if the dynamism were removed there would remain neither space nor the appearance of space. If this were true, and it is worthy of serious thought, then space is certainly finite, as in its totality, according to STANLEY's view, it would have to be regarded as a "phenomenon of the inner and finite life of the infinite."

It is believed that we may go a step further and unqualifiedly assert that *space is finite*, even denying its infinity as a "general mode of the activity of the whole." Yet it is transfinite in the sense that it transcends the comprehension of finite minds or processes. It is *finite* because it is in *manifestation*. Everything that is in manifestation is finite. The infinite is not in manifestation. Infinity has to be limited always to become manifest. The Deity has limited His being in order that there may be a manifested universe. All things, all appearances are finite; because they are phenomena connected with manifestation.

This question may be viewed from another standpoint. All things in manifestation or existence are polar in their constitution. For instance: there cannot be a "here" without a "there." There cannot be an "upper" without a "lower." Right is copolar with wrong; good is copolar with evil; night with day; manifestation with non-manifestation; truth with falsity; infinity with finity and so on, throughout the whole gamut of the pairs of opposites.

What is the logical inference? Space is paired with a lack of space. There cannot be what we call *space* without there being at the same time the possibility, at least, of the *lack of space or spacelessness*. This is a conclusion that is rigorously logical and incontrovertible.

But it has been urged that it is impossible for the mind to imagine a condition where there is no space. It even has been asserted that it is contrary to the constitution of the mind itself to imagine "no space." But whether imaginable or not has no effect whatever upon the validity of the conception. Neither, it is said, can we imagine a fourth dimension but the mind has come dangerously near to imagining it. The distance from excogitating upon, discussing and describing the properties of four-space to imagining it is not so great after all. Truly it is difficult indeed, it seems, to be able to describe a thing yet not be able to imagine or make a mental image of it. There is an evident fallacy here. Either the description of four-space is no description at all or it is a true delineation of an idealized construction which is well within the mind's powers of imagination. Indeed the question of imaginability is not determinative in itself; for what the mind may now be unable to imagine, because of its more or less nebulous character, and owing to its infancy may in the course of time be easily accomplished.

The universe is a compacted *plenum*. It is chock-full of mind, of life, of energy and matter. These four are basically one. They exist, of course, in varying degrees of tenuity and intensity and answer to a wide range of vibrations. Together, in their manifestation of action and interaction, in their *dynamic appearance*, if you please, they constitute space. If these were removed with all that their existence implies there would result a condition of spacelessness in which no one of the appearances which we now perceive would be possible. Even sheer extensity would be non-existent. All scope of motility would be lacking. Dimension, coördinates, direction, space-relations—all would be impossible.

A straight line is an ideal construction of the mind. It does not exist in nature. It can never be actualized in the phenomenal universe. Between the ideal and the real, or actual, there is a kosmic chasm. It broadens or narrows according as the phenomenal appearance approaches or recedes from the ideal. What, therefore, can be postulated of the one will not apply with equal force to the other. They are not congruent and can never be in the actualized universe. The moment the actual becomes identified with the ideal it ceases to be the actual. The universe does not exist as *pure form*, neither does space. As purely formal constructions of the intellect these can have no perceptible existence. The phenomenal or sensible may not be judged by exactly the same standard as the formal. The phenomenal or sensible represents things as they appear to the senses, or, so far as the actualized universe is concerned, *as they really are*. The *formal* represents things as they are made to appear by the mind. It cannot be actualized. It may be said that the purely formal is the limit of evolution. The phenomenal may approach the ideal as a limit, but can never become fully congruent with it. *The difference between the ideal and the actual is a dynamic one*; it is by virtue of this difference that the universe is held in manifestation. Evolution is the decrement of this difference between the purely formal and the actual. So long then as a kosmic differential is maintained the phenomenal continues to be manifest: when it is finally reduced to nothing it goes out of manifestation. The phenomenal is finite; the ideal infinite.

Wherefore, it is undoubtedly improper to refer to space as being infinite. The term really is inapplicable. Transfinity is much better and more accurate. Space is transfinite because its scope is greater than any finite scope of motility can encompass, because it exceeds finite comprehensibility.

RIEMANN's notion that space is limited gains weight in the light of the foregoing considerations. But he could not conceive of the limitability and unboundedness of space as such in its pure essence; but was compelled, by his own limitations, to make an idealized construction in which he could actualize his conception. And for real, dynamic space, he substituted his ideal construction and proceeded upon that basis. And of course, his view while it had no reference to perceptual space nevertheless possessed an illative relation thereto and should be recognized as construable in that light.

The process of squaring the circle recognized as a geometric impossibility is significant of the fluxional nature of the universal residuum perpetually maintained between the archetypal and the manifested kosmos. It seems that there is a profound truth embodied in this problem. There is a lesson that may be learned by mathematicians, philosophers, scientists and thinkers in general. There is an element of eternal necessity and universality about it which is truly symbolic of the finity of the universe and the infinity of the archetypal. Just as a square or a series of polygonal figures inscribed in a circle cannot be made to coincide exactly with the circle so cannot the actual be made to coincide with the ideal. The circumference of the circle is the unapproachable limit of inscribed squares. If it were possible so to multiply squares thus inscribed that a figure coincident with the circumference of a circle might be constructed, such a figure would not be a square but a circle. The manifested universe is like that—the process of inscribing squares within a circle. It is ever *becoming, evolving, developing*, but never quite attains. Infinity is a process. But no single stage in that process is infinite. Each is finite and their totality makes the infinity of the process. The universe manifested to the senses or the intellect is finite.

"Space," says PAUL CARUS, "is the possibility of motion in all directions."[14] To be sure, it is admitted that space offers opportunity for motion in all directions. But is space this opportunity of motility? Or is possibility of motion space? The possibility of motion must rest in the thing that moves. It implies a potency in the moving entity, not in space. If it is meant that space is the potency that resides in the moving element it is still more difficult to understand the connotation. But even granting this view, are we not compelled to recognize the dynamism of space as a necessary inference? Another definition which CARUS gives is that space is a "*pure form of extension*." If it be granted that space is a pure form of extension we should have to conclude that it has no actual existence; for *pure form* does not exist except as an idealized construction. It cannot be found in nature. Pure form is *ideal*. Impure or natural form is actual. Therefore the space in which we live and in which the universe exists cannot be a "pure form" because life cannot exist in the purely formal. It is useless to talk about space as mere form so long as it maintains life. The difficulty which this phase of the question presents is another evidence of the inadequacy of our definitions.

It is also found to be impossible to concur in CARUS' conception of knowledge *a priori*. His notion of the *a priori* varies somewhat from the Kantian view. He defines it as an "idealized construction," the "mind made," "abstract thought," and places it in the same category as a concept. This is undoubtedly born of his desire to get rid of KANT's "innate ideas" which seem to be distasteful to him. But in doing so it appears that the real *a priori* has been overlooked. Let us examine for a moment this important question. The *a posteriori* connotates all knowledge gained through the senses, or sense experience. All knowledge therefore whose origin can be traced to the senses is knowledge *a posteriori*. Now, knowledge *a priori* should be just the opposite of this. It should indicate such knowledge as that which does not have its origin in the senses, or which is not dependent upon the ordinary avenues of sense-experience. Abstract thought is as truly experience as smelling, seeing or hearing. It is by traversing its scope of motility that the mind finds out what the norms of logic are. It could not remain quiescent and discover them. It has to be active, examining, comparing and judging. Almost the entire range of thought, its entire scope, is characterized by the *a posterioristic method*. In fact, all thought is *a posterioristic*. Despite the fact that, in thinking in the abstract, it is necessary mentally to remove all elements of concreteness, all materiality and all actuality, the conclusions reached have to be referred to the standards maintained by the actual, the concrete and the material. Then we do not start with the abstract in our thinking. We begin with the concrete and by mentally removing all physical qualities arrive at the abstract.

The mind has a constitution. It acts in a given way because it is its nature so to act. Not because it has learned to act in that manner. It performs certain functions intuitively without previous instruction or experience for the same reason that water dampens or heat warms. It is natural for it to do so. This naturalness, this performance of function without being taught or without experience constitute the principle of *apriority* in the mind. *Aprioriness* is a principle of mind partaking of the very nature and essence of mind. It is the very mainspring of mentality. Perception and conception are processes which the mind performs intuitively. The mind perceives and conceives because it is impossible for the normal mind to do otherwise. We take a view upon a given question; we assume certain mental attitudes of affirmation, negation or indifference because we have learned to do so by virtue of the tuitional capability of mind. These describe the *a posteriori*. That is, all knowledge obtained as a result of voluntary mental processes constitutes the mass of knowledge *a posteriori*. The *a priori* is what the mind is by nature: the *a posteriori* is what the mind becomes. It is the mind-content.

The *a priori* is not a mental construction; it is an essential principle of mind. It should not be identified with the "purely formal," as is done by PAUL CARUS:[15]

He says:

> "The *a priori* is identical with the purely formal which originates in our mind by abstraction. When we limit our attention to the purely relational, dropping all other features out of sight, we produce a field of abstraction in which we can construct purely formal combination, such as numbers, or the ideas of types and species. Thus we create a world of pure thought which has the advantage of being applicable to any purely formal consideration and we work out systems of numbers which, when counting, we can use as standards of reference for our experience in practical life."

Thus CARUS definitely links up the *a priori* to a factor which is nothing more nor less than a mental by-product. For such is the category in which would be placed both the process of abstraction and its results. It is therefore exceedingly difficult to understand why so cursory a consideration should have been given to the principle of *apriority* than which no other element of mind is more essentially a part of the mind itself.

The formal is symbolic. It signifies an informing quantity. Pure form itself is but a negation of that which formerly filled it. Then, too, the formal is purely artificial because it is a mental construction. Essentially there is as much difference between the purely formal and the *a priori* as between creator and creature, as between potter and clay. The one is the builder, the other is the material; the one the knower and the other the known. Thus, the only reason that the formal is found to be answerable to the *a priori* at all is due to the fact that it is construable only upon the basis of the *a priori*. But being so is not sufficient warrant for its identification with the *a priori*. The formal merely represents the totality of possibilities in the universe as viewed by the mind; but as the number of possibilities open to the mind is, on account of its nature and purpose limited, it is not to be supposed that it (the mind) shall measure up to all the possibilities offered by the formal. Moreover, it is certain that no sane mind cherishes the hope that there shall ever be found in the universe of life and form a congruence for all of the possibilities held out by the purely formal.

As an eternal principle of mind, the *a priori* is in agreement with the divine mind of the kosmos. In its *aposteriority* the mind is of diverse tendencies, qualities and characteristics. Apriorily, it acts in unison with the eternal purpose of life and the universal mind. In its aposteriority, it often goes awry. In its *apriority* it can never be insane; insanity is a symptom of the morbid *a posteriori*.

The mind in man acts the same as mind in the vegetal and lower animal kingdoms. Metabolism and katabolism, indeed all cell-activity, are *a priori* performances of the mind. Growth and all its phenomena, the cyclicism of natural processes, and every activity connected therewith belong to the category of the *a priori*. Cells multiply, divide, build up and tear down tissues and they do it intuitively. Most certainly these functions are performed without any assistance from the intellect. All the myriad activities in nature with which the intellect in man has not the slightest concern, truly acting in accord with some primordial impetus, are activities *a priori*.

Now what is the attitude of the intellect, in the light of the *a priori*, towards space and the question of dimensionality? It is evident that no matter what this attitude may be it is in agreement with the constitution of things and of the universe. And if so, it is right and without illusion. It is also evident that whatever notion *a posteriori* the intellect may entertain with respect to these questions is unavoidably liable to the illusionary drawbacks common to conclusions based upon limited experience. The geometric view of space belongs to the category of the *a posteriori*. Hence it is subject to the usual imposition of error.

Tersely stated, KANT's view of space is that it is a form of intuition, a form *a priori*, a transcendental form. As such he considered it to be a native form of perception not belonging to the category of sense-deliveries. Accordingly, space is a form of intuition arising out of and inhering in the constitution of mind. It is a notion which constitutes the universal and eternal prerequisite of mind and is, therefore, intrinsically necessary to all phases of mentation. Now, this being true just what may be said to be the relation of dimensionality to this *a priori* form of space which is found to exist in the mind as an eternal aspect of its nature? Does the mind intuitively measure its contents or its operations by the empirical standard of space-measurement known as dimension? Is the attitude of the mind towards the objectively real one of discrimination *a priori* as to the direction or dimension in which a percept may originate? In other words, does the mind habitually and intuitively refer its data to a system of coördinates for final determination? There is no other answer but that the mind makes no such reference and is dependent upon no kind of coördinate system in any of its operations *a priori*. As a form of intuition, the space notion is present in the mind as a scope of existence, of motility, of being and of sheer roominess. The notion of direction or dimension, being an artificial construction, does not enter into this form of intuition at all. It is only when the mind comes to elaborate upon its perceptive performances and possibilities that the questions of relations, positions and directions arise. But this latter is a matter separate and distinct from the state of awareness which embodies the notion of space.

Dimension is an arbitrary norm constructed by the mind for the determination of various positions in space. It is an accident or by-product of the process of elaborative cognition, a convenient and appropriate means of measurement for objects in space and their space-relations. But it is no more *a priori* than a foot rule or a square. But being purely an empirical product it may be said to be an aspect of psychogenesis because it relates to the evolutionary aspect of mind. The assumption may therefore be allowed that the mind may, in the course of its evolution, find it convenient and appropriate to devise an additional ordinate or dimension to satisfy the necessities of its more complex ramifications into the nature of things and to determine their greatly increased space-relations. It may be even possible for the mind to function normally in a space of four dimensions. But this would simply be a new adjustment, not a change in the essential nature of mind. It would be like the series of adjustments to environments which man has made in the onward movement of civilization. There has been no serious change in the manhood *per se* of man. That has remained the same; there has been merely a complication of environmental influences.

Similarly, in the acquisition of four-dimensional powers, granting that such an acquisition is possible, there is nothing to be added to the *aprioriness* of mind *itself*. Is it not, therefore, logical to assume that the discovery of a fourth coördinate and the consequent conceptualization of the same, point to the development in the mind of a greatly extended faculty, more keenly penetrative powers of cognition and a further diversification of its environments than it has hitherto enjoyed? Indeed, it seems so.

CHAPTER V

The Fourth Dimension

The Ideal and the Representative Nature of Objects in the Sensible World—The Psychic Fluxional the Basis of Mental Differences—Natural and Artificial Symbols—Use of Analogies to Prove the Existence of a Fourth Dimension—The Generation of a Hypercube or Tesseract—Possibilities in the World of the Fourth Dimension—Some Logical Difficulties Inhering in the Four-Space Conception—The Fallacy of the Plane-Rotation Hypothesis—C. H. Hinton and Major Ellis on the Fourth Dimension.

The world of mathesis is truly a marvelous domain. Vast are its possibilities and vaster still its sweep of conceivability. It is the kingdom of the mind where, in regal freedom, it may perform feats which it is impossible to actualize in the phenomenal universe. In fact, there is no necessity to consider the limitations imposed by the actualities of the sensuous world. Logic is the architect of this region, and for it there is no limit to the admissibility of hypotheses. These may be multiplied at will, and legitimately so. The chief error lies in the attempt to make them appear as actual facts of the physical world.

Mathematicians, speculating upon the possibilities of mathetic constructions and forgetting the necessary distinctions which should be recognized as differentiating the two worlds, in their enthusiasm have been led into the error of postulating as qualities of the phenomenal world the characteristics of the conceptual. Accordingly, a great deal of confusion as to the proper limits and restrictions of these conceptions has arisen and there still may be found those who are enthusiastically endeavoring to push the actualities of the physical over into the conceptual. But in assuming any attitude towards mathetic propositions, especially with a view to demonstrating their actuality, very careful discrimination as to the essential qualities and their connotations should be made. Hence, before taking up a brief study of the fourth dimension proper, it is deemed fitting to indicate some of the fundamental distinctions which every student of these questions should be able to make with reference to the data which he meets.

All objects of the sensible world have both an essential or ideal nature and a representative or sensuous nature. That is, they may be studied from the standpoint of the ideal as well as the sensuous. The representative nature is that which we recognize as the mode of appearance to our senses which, as KANT held, is not the essential or ideal character of the thing itself. For there is quite as much difference between the sensuous percept and the real thing itself as between an object and its shadow. In fact, a concept viewed in this light, may be seen to have all the characteristics of an ordinary shadow; for instance, the shadow of a tree. View it as the sun is rising; it will then be seen to appear very much elongated, becoming less in length and more distinct in outline as the sun rises to a position directly overhead. The elongation may again be seen when the sun is setting. Throughout the day as the sun assumes different angles with reference to the tree the proportions and definiteness of the shadow vary accordingly. Thus the angularity of the sun, the intensity and fullness of the light and the shape and size of the tree operate to determine the character of the shadow.

Much the same thing is true of a sensuous representation. If we examine carefully our ideas of geometric quantities and magnitudes, it will be found that the concepts themselves are not identical with the objects of the physical world, but mere mental shadows of them. The angularity of consciousness, or the distinctness of one's state of awareness, being analogous to similar attitudes in the solar influence are the main determinants of the character of the mental shadow or concept. Wherefore mathematical "spaces" or magnitudes are not sensuous things and have therefore no more real existence than a shadow, and strictly speaking not as much; for a shadow may be seen, while such magnitudes can only be conceived. It may be urged that since we can conceive of such things they must have existence of some kind. And so they have, but it is an existence of a different kind from that which we recognize as belonging to things in the sensible world. They have a conceptual existence, but not a sensuous one. Therein lies the great difference.

To be sure, a shadow is a more or less true representation of the thing to which it pertains. That this is true can be established empirically. Similarly, the degree of congruity between objects and concepts likewise may be determined. If this were not true we should be very much disappointed with what we find in the phenomenal world and could never be quite sure that the mentograph existing in our minds was a faithful representation of the thing which we might be examining. But really the foundation for such a disappointment is present in every concept, every percept with which the mind deals. This disappointment, although in actual experience is reduced to an almost negligible quantity, is due to the failure of sensuous objects to conform wholly to the specific details of the mental shadow or mentograph. This lack of congruence between the mental picture and the object itself is necessary for obvious reasons. It is markedly observable in the early efforts of a child in learning distances, weights, resistances, temperatures and the like. No inconsiderable time is required for the child to be able correctly to harmonize his sense-deliveries with actual conditions. Otherwise, the child would never make any of the ludicrous mistakes of judgment of which it is guilty when trying to get its bearings in the world of the senses. In the course of time the child gradually learns by experience that certain things are true of objects, distances, temperatures, resistances, etc., and that certain things are not true of them. He learns these things by actually contacting various objects. He is then competent to render correct judgments, within certain limits, as to the conditions which he finds in the sensible world. And the allowances, equations and corrections which his motor, sensory and psychic mechanisms learn to make in childhood serve for all subsequent time. And this is important to remember; for the mature mind is apt to forget or overlook the adaptations which the child-mind has made in its growth.

If there were no such differences between the concept and the thing itself, actual physical contact would not be necessary. For one could rely wholly upon the sense-deliveries and each sense might operate entirely independently of all the others as there would be no necessity to correct the delivery of one by those of the others. This, of course, raises the question as to the necessity of sense-experience at all under conditions where there would be no disparity between the thing itself and the ideal representation of it in the mind. The absence of this variable quantity would open to the mind the possibility of really knowing the essential nature of objects in the phenomenal world, a condition of affairs which is admittedly now without the range of the powers of the mind.

At any rate, the essential "thingness" of objects can never be comprehended by the mind until the diminution of this disparity between the object of sense and the mental picture of it which exists in the consciousness has proceeded to such a limit as either completely to have obliterated it or to such an extent that the psychic fluxion is so slight as not to matter.

It is believed that the results of mental evolution, as the mind approaches the transfinite as a limit, will operate to minimize the fluxional quantity which subsists between all objects of sense and their ideal representation as data of consciousness. The conclusion that the mind of early men who lived hundreds of thousands and perhaps millions of years ago on this planet consumed a much longer time in learning the adjustments between the objects which it contacted in the sensuous world and the elementary representations which were registered in its youthful consciousness than is to-day required for similar processes seems to be demanded, and substantiated as well, by what is known of the phyletic development of the mind in the human race.

In view of the above, it is thought that the duration of such simple mental processes served not only to prolong the physical life of the man of those early days, but may also account for the puerility and incapacity of the mind at that stage. Not that the slow mental processes were active causative agencies in lengthening the life of man, but that they together with the crass physicality of man necessitated a longer physical life. This, perhaps in a larger sense than any other consideration, accounts for the fundamental discrepancies in the mind of the primitive man in comparison with the efficiency of the mind of the present-day man. In view of the potential character of mind and in the light of the well graduated scale of its accomplishments, it is undoubtedly safe to conclude that the quality of mental capacities is proportional to the psychic fluxional which may exist at any time between the ideal and the essential or real. Mental differences and potentialities in general may be due to the magnitude of the psychic fluxional or differential that exists between the conceptual and the perceptual universe. In some minds it may be greater than in others. The chasm between things-in-themselves and the mental notion pertaining thereto may vary in a direct ratio to the individual mind's place in psychogenesis, and therefore, be the key to all mental differences in this respect.

Most certain it is that there may be marked fluctuations in the judicial approach of minds towards any psychic end. In other words, there is not only a fluxional or differential between the object and its representation, but also a differential between the approach of one mind and another in the judicial determination of notions concerning ideas. In this way, differences of opinions as to the right and wrong of judgments arise. Indeed, there seem to be zones of affinity for minds of similar characteristics, or minds that have the same degree of differential; so that, in choosing among the many possible judgments predicable upon a species of data, all those minds having the same degree of psychic differential discover a special affinity or agreement among themselves. Hence, we have cults, schools of thought, and various other sectional bodies that find a basis of agreement for their operations in this way. The outcome of this remarkable intellectual phenomenon is that there are as many different kinds of judgments as there are zones of affinity among minds. Various systems of philosophy owe their existence to these considerations, and the considerations themselves flow from the fact that all intellectual operations are essentially superficial; because there is no means by which they may penetrate to the steady flowing stream of reality which pervades and sustains objects in the sensible world.

In view, therefore, of the foregoing and with special reference to geometric constructions, it is necessary in approaching a study of the four-space that it be understood at the outset that the fourth dimension can neither be actualized nor made objectively possible even in the slightest degree in the perceptual world; because it belongs to the world of pure thought and exists there as an "extra personal affair," separate and distinct from the world of the senses.

As says SIMON NEWCOMB:[16]

> "The experience of the race and all the refinements of modern science may be regarded as showing quite conclusively that, within the limits

of our experience, there is no motion of material masses, in the direction of a fourth dimension, no physical agency which we can assume to have its origin in regions to which matter cannot move, when it has three degrees of freedom."

There is, however, no logical objection to the study of the fourth dimension as a purely hypothetical question, if by pursuit of the same an improvement of methods of research and of the outlook upon the field of the actual may be gained. Hence, it is with this attitude of mind that we approach the consideration of the fourth dimension.

Various efforts have been made to render the conception of a fourth dimension of space thinkable. The student of space has reasoned: "We say that there are three dimensions of space. Why should we stop here? May there not be spaces of four dimensions and more?" Or he has said: "If 'A' may represent the side of a square, A^2 its area, and A^3 the volume of a cube with edge equal to A; what may A^4, A^5 or A^{nth} represent in our space? The conclusion, with respect to the quantity A^4, has been that it should represent a space of four dimensions."

Algebraic quantities, however, represent neither objects in space nor space qualities except in a purely conventional manner. All efforts to justify the objective existence of a fourth dimension based upon such reasoning will, therefore, fail; because the basis of such arguments is itself faulty. In the sentence: "The man loves his bottle," the thing meant is not the bottle, but what the bottle contains. For the purpose of the figure the bottle signifies its contents. There is no more real connection between the bottle and what it contains than between any word and the object for which it stands. Words are said to be symbols of ideas. But they are not natural symbols; they are conventional symbols, made for the purpose of cataloguing, indexing and systematizing our knowledge. Words can be divorced from ideas and objects, or rather have never had any real connection with them. There are two classes of natural symbols, namely; *objects* and *ideas*. These, objects and ideas, symbolize realities. Realities are imperceptible and incomprehensible to the intellect which has aptitude only for a slight comprehension of the symbols of realities. For instance, a tree is a natural symbol. It represents an actuality which is imperceptible to the intellect. The intellect can deal only with the sensible symbol. It is a natural symbol;

because it is possible directly to trace a living connection between the tree and the *tree-reality*. That is, it would be possible so to trace out the vital connection between the tree and its reality if the intellect had aptitude for such tracery. But, in reality, since it has no such aptitude, it remains for the work of that higher faculty than the intellect which recognizes both the connection and the intellect's inability to trace it. Further, an object is called a natural symbol because it is the bridge between sensuous representation and reality. It is as if one could begin at the surface of an object and by a subtle process of elimination and excortication arrive at the heart of the universum of reality. No such consummation may be reached by dealing with words which have merely an artificial relationship with the objects which they signify. Again, ideas, that is, ideas that are universal in application and have their roots in the great ocean of reality, are natural symbols; because if it were possible to handle an idea with the physical hands it would be possible to arrive at the heart of that which it symbolized without ever losing our connection with the idea itself. In other words, ideas and objects, unlike words, can never be divorced from that which they symbolize. Both, being of the same class, are the opposite poles of realities. This then is the difference between natural symbols and artificial symbols— that a natural symbol, such as objects and ideas, is copolar with reality whereas an artificial symbol, such as words, geometric constructions and the like not only lacks this copolarity but is itself a symbol of natural symbols.

It is, therefore, inconceivable that because the algebraic quantity A^3 has been arbitrarily decreed to be a representation of the volume of a cube, every such quantity in the algebraic series shall actually represent some object or set of objects in the physical world. Even if it be granted that such may be the case, is it not certain that there is a limit to things in the objective universe? Yet there may not be any limit to algebraic or mathematical determinations. The material universe is limited and conditioned; the world of mathesis is unlimited and unconditioned save by its own limitations and conditions. It is irrational to expect that physical phenomena shall justify all mathematical predicates.

There is perhaps no single mathematical desideratum or consideration which may be said to be the natural symbolism of realities; for the whole of mathematical conclusions is a mass of artificial and arbitrary but concordant symbols of the crasser or nether pole of the antipodes of realism. It is exceedingly dangerous, therefore, to predicate upon such a far-fetched symbolism as mathematics furnishes anything purporting to deal with ultimate realities. And those who insist upon doing so are either blind themselves to these limitations or are madly endeavoring to befog the minds of others who are dependent upon them for leadership in questions of mathematical import.

Analogies have been unsparingly used in efforts to popularize the four-space conception and much of the violence which has been done to the notion is due to this vagary. The mathematical publicist, in trying to give a mental picture of the fourth dimension, examines the appearances of three dimensional beings as they might appear to a two dimensional being or *duodim*. He imagines a race of beings endowed with all the human faculties except that they live in a land of but two dimensions—length and breadth. He thinks of them as shadows of three dimensional beings to whom there are no such conceptions as "up" and "down." They can see nothing nor sense anything in any way that is without their plane. They can move in any direction within the plane in which they live, but can have no idea of any movement that might carry them without that plane. A house for such beings might be simply a series of rectangles. One of them might be as safe behind a line as a *tridim* or three dimensional being would be behind a stone wall. A bank safe for the *unodim* would be a mere circle. A *duodim* in any two dimensional prison might be rescued by a tridim without the opening of doors or the breaking of walls. An action of a *tridim* performed so as to contact their plane would be to them a miracle, absolutely unaccountable upon the basis of any known fact to the *unodim* or *duodim*. A *tridim* might go into a house where lived a family of *duodims*, appear and disappear without being detected or its ever being discovered how he accomplished such a marvelous feat. Our miracles, after the same fashion, are said to be the antics of some four dimensional being who has similar access to our three dimensional world and whose actions are similarly inexplicable to us. So the analogies have been multiplied. But the temptation to apply the consequences of such reasoning to actual three-space conditions has been so great that many have yielded to it and have consequently sought actually to explain physical phenomena upon the basis of the fourth dimension.

The utilitarian side of the question of hyperspace has not been neglected either. And so, early in the development of the hypothesis and its various connotations, the attention of investigators was turned to this aspect of the inquiry. Strange possibilities were revealed as a result. For instance, it was found that an expert fourth dimensional operator is possessed of extraordinary advantages over ordinary tridimensional beings. Operating from his mysterious hiding place in hyperspace, he could easily appear and disappear in so mysterious a manner that even the most strongly sealed chests of treasures would be easily and entirely at his disposal. No city police, Scotland Yard detective nor gendarme could have any terrors for him. Drs. Jekyll and Messrs. Hyde might abound everywhere without fear of detection. Objects as well as persons might be made to pass into or out of closed rooms "without penetrating the walls," thus making escape easy for the imprisoned. No tridimensional state, condition or system of arrangements, accordingly, would be safe from the ravages of evilly inclined four dimensional entities. Objects that now are limited to a point or line rotation could in the fourth dimension rotate about a plane and thus further increase the perplexities of our engineering and mechanical problems; four lines could be erected perpendicular to each other whereas in three space only three such lines can be erected; the right hand could be maneuvered into the fourth dimension and be recovered as a left hand; the mysteries of growth, decay and death would find a satisfactory explanation on the basis of the fourth dimensional hypothesis and many, if not all, of the perplexing problems of physiology, chemistry, physics, astronomy, anthropology and psychology would yield up their mysteries to the skill of the fourth dimensional operator. Marvelous possibilities these and much to be desired! But the most remarkable thing about these so-called possibilities is their impossibility. It is this kind of erratic reasoning that has brought the conception of a fourth dimension into general disrepute with the popular mind. It is to be regretted, too, for the notion is a perfectly legitimate one in the domain of mathesis where it originated and rightly belongs.

It is not to be wondered at that metageometricians and others should at first surmise that, in the four-space, they had found the key to the deep mysteries of nature in all branches of inquiry. For so vast was the domain and so marvelous were the possibilities which the new movement revealed that it was to be expected that those who were privileged to get the first glimpses thereof would not be able to realize fully their significance. But the stound of their minds and the attendant magnification of the elements which they discovered were but incidents in the larger and more comprehensive process of adjustment to the great outstanding facts of psychogenesis which is only faintly foreshadowed in the so-called hyperdimensional. The whole scope of inquiry connected with hyperspace is not an end in itself. It is merely a means to an end. And that is the preparation of the human mind for the inborning of a new faculty and consequently more largely extended powers of cognition. Metageometrical discoveries are therefore the excrescences of a deeper, more significant world process of mental unfoldment. They belong to the matutinal phenomena incident to this new stage of mental evolution. All such investigations are but the preliminary exercises which give birth to new tendencies which are destined to flower forth into additional faculties and capacities. So that it is well that the evolutionary aspect of the question be not overlooked; for there is danger of this on account of the magnitude and kosmic importance of its scope of motility.

A geometric line is said to be a space of one dimension. A plane is a space of two dimensions and a cube, a space of three dimensions. In figure 7 below, the line *ab* is said to be one dimensional; because only one coördinate is necessary to locate a point-position in it. The plane, *abcd*, figure 8, is said to be two dimensional because two coördinates, *ab* and *db* are required to locate a point, as the point *b*. The cube *abcdefgh*, figure 9, is said to be tridimensional, because, in order to locate the point *b*, for instance, it is necessary to have three coördinates, *ab*, *bc* and *gb*. The tesseract is said to be four dimensional, because, in order to locate the point *b*, in the tesseract, it is necessary to have four coördinates, *ab*, *bc*, *bb'* and *h'b*, figure 10.

FIG. 7.

FIG. 8.

It will be noted that in figures 8, 9 and 10, the element of perpendicularity enters as a necessary determination. In figure 8, the lines *ab* and *bd* are perpendicular to each other. Similarly, in Fig. 10, lines *ab*, *bc*, *bb'* and *h'b* are perpendicular to one another. That is, at their intersections, they make right angles. Similarly, figures representing any number of dimensions may be constructed.

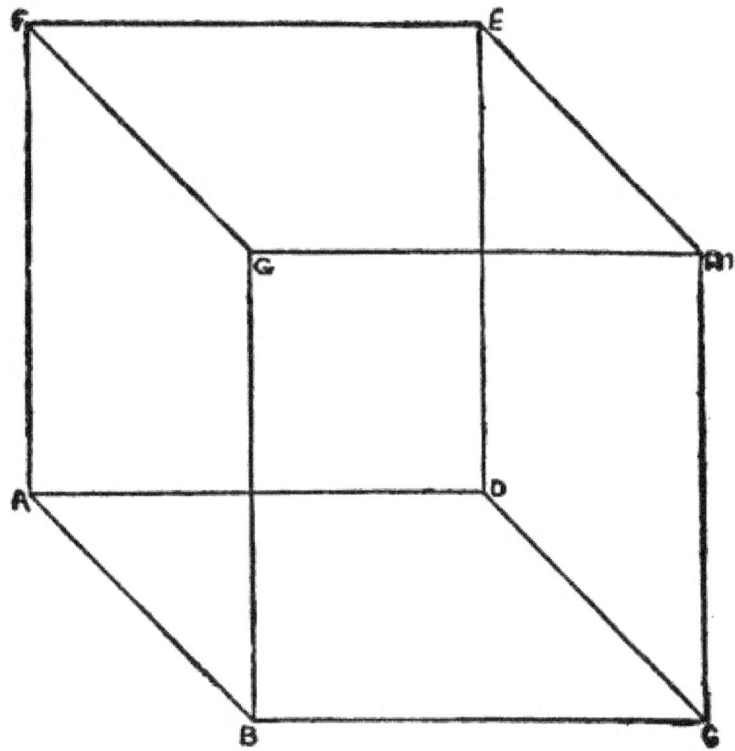

FIG. 9.

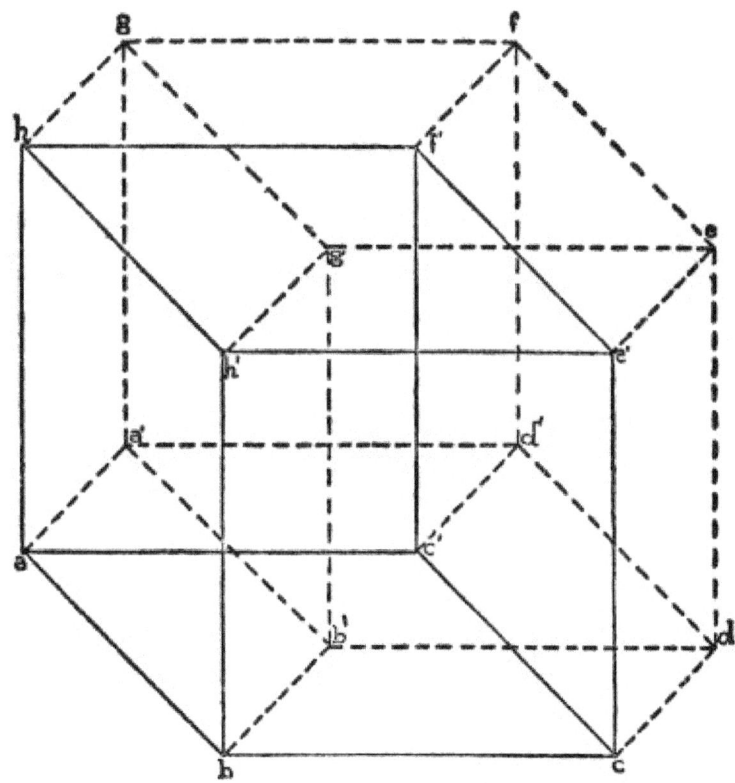

FIG. 10.

The line *ab* represents a one-space. An entity living in a one space is called a "unodim." The plane, *abcd*, represents a two-space, and entities living in such a space are called *duodims*. The cube, *abcdefgh*, represents a three-space and entities inhabiting such a space are called *tridims*. Figure 10 represents a four-space, and its inhabitants are called *quartodims*. Each of the above-mentioned spaces is said to have certain limitations peculiar to itself.

The fourth dimension is said to lie in a direction at right angles to each of our three-space directions. This, of course, gives rise to the possibility of generating a new kind of volume, the hypervolume. The hypercube or tesseract is described by moving the generating cube in the direction in which the fourth dimension extends. For instance, if the cube, Fig. 9, were moved in a direction at right angles to each of its sides a distance equal to one of its sides, a figure of four dimensions, the tesseract, would result.

The initial cube, *abcc'e'fhh'*, when moved in a direction at right angles to each of its faces, generates the hypercube, Fig. 10. The lines, *aa'*, *bb'*, *cc'*, *dd'*, *ee'*, *ff'*, *gg'*, *hh'*, are assumed to be perpendicular to the lines meeting at the points, *a*, *b*, *c*, *d*, *e*, *f*, *g*, *h*. Hence *a'b'*, *b'd*, *dd'*, *d'a'*, *ef*, *fg*, *gg'*, *g'e*, represent the final cube resulting from the hyperspace movement. Counting the number of cubes that compose the hypercube we find that there are eight. The generating cube, *abcc'e'f'hh'*, and the final cube, *a'b'*, *b'd*, *dd'*, *d'a'*, *ef*, *fg*, *gg'*, *g'e*, make two cubes; and each face generates a cube making eight in all. A tesseract, therefore, is a figure bounded by eight cubes.

To find the different elements of a tesseract, the following rules will apply:

1. *To find the number of lines*: Multiply the number of lines in the generating cube by two, and add a line for each point or corner in it. E.g., 2 × 12 = 24 + 8 = 32.

2. *To find the number of planes, faces or squares*: Multiply the number of planes in the generating cube by 2 and add a plane for each line in it. E.g., 2 × 6 + 12 = 24.

3. *To find the number of cubes in a hypercube*: Multiply the number of cubes in the generating cube, one, by two and add a cube for each plane in it. E.g., 2 × 1 + 6 = 8.

4. *To find the number of points or corners*: Multiply the number of corners in the generating cube by 2. E.g., 2 × 8 = 16.

In a plane there may be three points each equally distant from one another. These may be joined, forming an equilateral triangle in which there are three vertices or points, three lines or sides and one surface.

In three-space there may be four points each equidistant from the others. At the vertices of a regular tetrahedron may be found such points. The tetrahedron has four points, one at each vertex, 6 lines and 4 equilateral triangles, as in Fig. 11.

In four-space, we have 5 points each equidistant from all the rest, giving the hypertetrahedron. This four dimensional figure may be generated by moving the tetrahedron in the direction of the fourth dimension, as in Fig. 12. If a plane be passed through each of the six edges of the tetrahedron and the new vertex there will be six new planes or faces, making 10 in all, counting the original four. From the new vertex there is also a tetrahedron resting upon each base of the original tetrahedron so that there are five tetrahedra in all. *A hypertetrahedron is a four-dimensional figure consisting of five tetrahedra, ten faces, 10 lines and 5 points.*

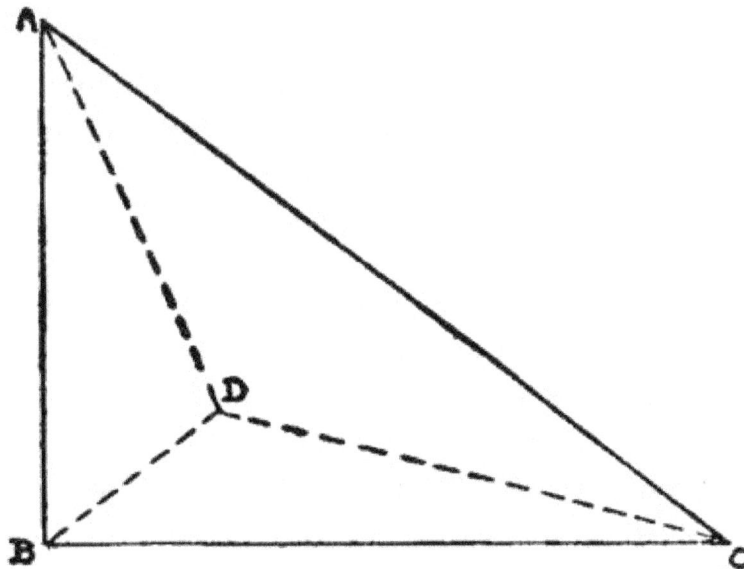

FIG. 11.

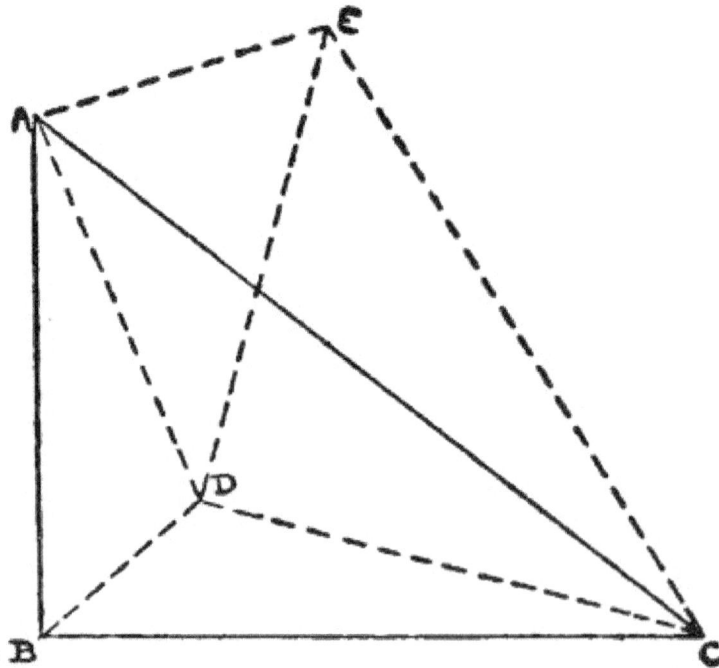

FIG. 12.

PAUL CARUS[17] suggests the use of mirrors so arranged that they give eight representations of a cube when placed at their point of intersection. He says:

"If we build up three mirrors at right angles and place any object in the intersecting corner we shall see the object not once, but eight times. The body is reflected below and the object thus doubled is mirrored not only on both upright sides but in addition in the corner beyond, appearing in either of the upright mirrors coincidingly in the same place. Thus the total multiplication of our tridimensional boundaries of a four dimensional complex is rendered eight-fold.

"We must now bear in mind that this representation of a fourth dimension suffers from all the faults of the analogous figure of a cube in two dimensional space. The several figures are not eight independent bodies but are mere boundaries and the four dimensional space is conditioned by their interrelation. It is that unrepresentable something which they inclose, or in other words, of which they are assumed to be boundaries. If we were four dimensional beings we could naturally and easily enter into the mirrored space and transfer tridimensional bodies or parts of them into those other objects reflected here in the mirrors representing the boundaries of the four dimensional object. While thus on the one hand the mirrored pictures would be as real as the original object, they would not take up the space of our three dimensions, and in this respect, our method of representing the fourth dimension by mirrors would be quite analogous to the cube pictured on a plane surface, for the space to which we (being limited to our tridimensional space-conception), would naturally relegate the seven additional mirrored images is unoccupied and if we should make the trial, we would find it empty."

The utility of such a representation as that which CARUS outlines in the above is granted, i.e., so far as the purpose which it serves in giving a general idea of what a four-space object might be imagined to be like, but the illustration does not demonstrate the existence of a fourth dimension. It only shows what might be if there were a four-space in which objects could exist and be examined. We, of course, have no right to assume that because it can be shown by analogous reasoning that certain characteristics of the fourth dimensional object can be represented in three-space the possible existence of such an object is thereby established. Not at all. For there is no imaginable condition of tridimensional mechanics in which an object may be said to have an objective existence similar to that represented by the mirrored cube.

But there are discrepancies in this representation which well might be considered. They have virtually the force of invalidating somewhat the conception which the analogy is designed to illustrate. For instance, in the case of the mirrored object placed at the point of intersection of the three mirrors built up at right angles to each other. Upon examination of such a construction it is found that the reflection of the object in the mirrors has not any perceptible connection with the object itself. And this, too, despite the fact that they are regarded as boundaries of the hypercube; especially is this true when it is noted that these reflections are called upon to play the part of real, palpable boundaries. If a fourth dimensional object were really like the mirror-representation it would be open to serious objections from all viewpoints. The replacement of any of the boundaries required in the analogy would necessarily mean the replacement of the hypercube itself. In other words, if the real cube be removed from its position at the intersection of the mirrors no reflection will be seen, and hence no boundaries and no hypercube. The analogy while admittedly possessing some slight value in the direction meant, is nevertheless valueless so far as a detailed representation is concerned. So the analogy falls down; but once again is the question raised as to whether the so-called fourth dimension can be established or proven at all upon purely mathematical grounds. It also emphasizes the necessity for a clearer conception of the meaning of dimension and space.

The logical difficulties which beset the hyperspace conception are dwelt upon at length by JAMES H. HYSLOP. He says:[18]

"The supposition that there are three dimensions instead of one, or that there are only three dimensions is purely arbitrary, though convenient for certain practical purposes. Here the supposition expresses only differences of directions from an assumed point. Thus what would be said to lie in a plane in one relation would lie in the third dimension in another. There is nothing to determine absolutely what is the first, second, or third dimension. If the plane horizontal to the sensorium be called plane dimension, the plane vertical to it will be called solid, or the third dimension, but a change of position will change the names of these dimensions without involving the slightest qualitative change or difference in meaning.

"Moreover, we usually select three lines or planes terminating vertically at the same point, the lines connecting the three surfaces of a cube with the same point, as the representative of what is meant by three dimensions, and reduce all other lines and planes to these. But interesting facts are observable here. 1. If the vertical relation between two lines be necessary for defining a dimension, then all other lines than the specified ones are either not in any dimension at all, or they are outside the three given dimensions. This is denied by all parties, which only shows that a vertical relation to other lines is not necessary to the determination of a dimension. 2. If lines outside the three vertically intersecting lines still lie in dimension or are reducible to the other dimensions they may lie in more than one dimension at the same time which after all is a fact. This only shows that qualitatively all three dimensions are the same and that any line outside of another can only represent a dimension in the sense of *direction* from a given point or line, and we are entitled to assume as many dimensions as we please, all within three dimensions.

"This mode of treatment shows the source of the illusion about the 'fourth dimension.' The term in its generic import denotes commensurable quality and denotes only one such quality, so that the property supposed to determine non-Euclidean geometry must be qualitatively different from this, if its figures involve the necessary qualitative differentiation from Euclidean mathematics. But this would shut out the idea of 'dimension' as its basis which is contrary to the supposition. On the other hand, the term has a specific meaning which as different qualitatively from the generic includes a right to use the generic term to describe them differentially, but if used only quantitatively, that is, to express direction as it, in fact, does in these cases, involves the admission of the actual, not a supposititious, existence of a fourth dimension which again is contrary to the supposition of the non-Euclidean geometry. Stated briefly, dimension as commensurable quality makes the existence of the fourth dimension a transcendental problem, but as mere direction, an empirical problem. And the last conception satisfies all the requirements of the case because it conforms to the purely quantitative differences which exist between Euclidean and non-Euclidean geometry as the very language about 'surfaces,' 'triangles,' etc., in spite of the prefix 'pseudo,' necessarily implies."

Thus it would seem that those who have been most diligent in constructing the hyperspace conception have been the least careful of the logical difficulties which beset the elaboration of their assumptions. Yet it sometimes requires the illogical, the absurd and the aberrant to bring us to a right conception of the truth, and when we come to a comparison of the two, truth and absurdity, we are the more surprised that error could have gained so great foothold in face of so overwhelming evidences to the contrary.

The entire situation is, accordingly, aptly set forth by HYSLOP when he says, continuing:

"There are either a confusion of the abstract with the concrete or of quantitative with qualitative logic, ... so that all discussion about a fourth dimension is simply an extended mass of equivocations turning upon the various meanings of the term 'dimension.' This when once discovered, either makes the controversy ridiculous or the claim for non-Euclidean properties a mere truism, but effectually explodes the logical claims for a new dimensional quality of space as a piece of mere jugglery in which the juggler is as badly deceived as his spectators. It simply forces mathematics to transcend its own functions as defined by its own advocates and to assume the prerogatives of metaphysics."

Shall we, therefore, assent to the imperialistic policy of mathematicians who would fain usurp the preserves of the metaphysician in order that they may exploit a superfoetated hypothesis? It is not believed that the harshness of HYSLOP's judgment in this respect is undeserved. It is, however, regretted that the notions of mathematicians have been so inchoate as to justify this rather caustic, though appropriate criticism. For it does appear that the moment the mathematician deserts the province of his restricted sphere of motility and enters the realm of the transcendental, that moment he loses his way and becomes an inexperienced mariner on an uncharted sea.

It is interesting to note that CASSIUS JACKSON KEYSER,[19] while recognizing the purely arbitrary character of the so-called dimensionality of space, nevertheless lends himself to the view that "if we think of the line as generating element we shall find that our space has four dimensions. That fact may be seen in various ways, as follows:

"A line is determined by any two of its points. Every line pierces every plane. By joining the points of one plane to all the points of another, all the lines of space are obtained. To determine a line, it is, then, enough to determine two of its points, one in the one plane and one in the other. For each of these determinations two data, as before explained, are necessary and sufficient. The position of the line is thus seen to depend upon four independent variables, and the four dimensionality of our space *in lines* is obvious."

Similarly he argues for the four dimensionality of space in spheres:

"We may view our space as an assemblage of its spheres. To distinguish a sphere from all other spheres, we need to know four and but four independent facts about it, as say, three that shall determine its center and one its size. Hence our space is four dimensional also in spheres. In circles, its dimensionality is six; in surfaces of second order (those that are pierced by a straight line in two points), nine; and so on ad infinitum."

The view taken by KEYSER is a typical one. It is the mathematical view and is characterized by a certain lack of restraint which is found to be peculiar to the whole scheme of thought relating to hyperspace. It is clear that the kind of space that will permit of such radical changes in its nature as to be at one time three dimensional, at another time four dimensional, then six, nine and even n-dimensional is not the kind of space in which the objective world is known to exist. Indeed, it is not the kind of space that really exists at all. In the first place, a line cannot generate perceptual space. Neither can a circle, nor a sphere nor any other geometrical construction. It is, therefore, not permissible, except mathematically, to view our space either as "an assemblage of its spheres," its circles or its surfaces; for obviously perceptual space is not a geometrical construction even though the intellect naturally finds inhering in it a sort of latent geometrism which is kosmical. For there is a wide difference between that kosmic order which is space and the finely elaborated abstraction which the geometer deceives himself into identifying with space. There is absolutely neither perceptible nor imperceptible means by which perceptual space in anywise can be affected by an act of will, ideation or movement. Just why mathematicians persist in vagarizing upon the generability of space by movement of lines, circles, planes, etc., is confessedly not easily understood especially when the natural outcome of such procedure is self-stultification. It is far better to recognize, as a guiding principle in all mathematical disquisitions respecting the nature of space that the possibilities found to inhere in an idealized construction cannot be objectified in kosmic, sensible space. The line of demarkation should be drawn once for all, and all metageometrical calculations and theories should be prefaced by the remark that: "if objective space were amenable to the peculiarities of an idealized construction such and such a result would be possible," or words to that effect. This mode of procedure would serve to clarify many if not all of the hyperspace conceptions for the non-mathematician as well as for the metageometricians themselves, especially those who are unwilling to recognize the utter impossibility of their constructions as applied to perceptual space. We should then cease to have the spectacle of otherwise well-demeanored men committing the error of trying to realize abstractions or abstractionizing realities. Herein is the crux of the whole matter, that mathematicians, rather than be content with realities as they find them in the kosmos, should seek to reduce them to abstractions, or, on the other hand, make their abstractions appear to be realities.

KEYSER proceeds to show how the concept of the generability of hyperspace may be conceived by beginning with the point, moving it in a direction without itself and generating a line; beginning with the line, treating it similarly, and generating a plane; taking the plane, moving it in a direction at right angles to itself and generating a cube; finally, using the cube as generating element and constructing a four-space figure, the tesseract. Now, as a matter of fact, a point being intangible cannot be moved in any direction neither can a point-portion of sensible space be removed. Nevertheless, we quite agree with him when he asserts:

> "Certainly there is naught of absurdity in supposing that *under suitable stimulation the human mind may, in the course of time, speedily develop a spatial intuition of four or more dimensions*." (The italics in the above quotation are ours.)

Here we have a tacit implication that the notion which geometers have heretofore designated as "dimension" really is a matter of consciousness, of intuition, and therefore, determinable only by the limitations of consciousness and the deliveries of our intuitive cognitions. As a more detailed discussion of this phase of the subject shall be entered into when we come to a consideration of Chapter VI on "Consciousness as the Norm of Space Determinations" further comment is deferred until then.

Now, as it appears certain that what geometers are accustomed to call "dimension" is both relative and interchangeable in meaning—the one becoming the other according as it is viewed—the conclusion very naturally follows that neither constructive nor symbolic geometry is based upon dimension as commensurable quality. The real basis of the non-Euclidean geometry is dimension as direction. For whatever else may be said of the fourth dimension so-called it is certainly unthinkable, even to the metageometricians, when it is absolved from direction although no specific direction can be assigned to it. It is agreed perhaps among all non-Euclidean publicists that the fourth dimension must lie in a "direction which is at right angles to all the three dimensions." But if they are asked how this direction may be ascertained or even imagined they are nonplused because they simply do not know. The difficulty in this connection seems to hinge about the question of identifying the conditions of the world of phantasy with those of the world of sense. There are distortions, ramifications, submersibles, duplex convolutions and other mathetic acrobatics which can be performed in the realm of the conceptual the execution of which could never be actualized in the objective world. Because these antics are possible in the premises of the mathematical imagination is scarce justification for the attempts at reproduction in an actualized and phenomenal universe.

One of the proudest boasts of the fourth dimensionist is that hyperspace offers the possibility of a new species of rotation, namely, *rotation about a plane*. He refers to the fact that in the so-called one-space, rotation can take place only about a point. For instance in Figure 7, the line *ab* represents a one-space in which rotation can take place only about one of the two points *a* and *b*. In Figure 8 which represents a two-space, rotation may take place about the line *ab* or the line *cd*, etc., or, in other words, the plane *abcd* can be rotated on the axial line *ab* in the direction of the third dimension. In tridimensional space only two kinds of rotation are possible, namely, rotation about a point and about a line. In the fourth dimension it is claimed that rotation can take place about a plane. For example, the cube in Figure 9, by manipulation in the direction of the fourth dimension, can be made to rotate about the side *abgf*.

A very ingenious argument is used to show how rotation about a plane is thinkable and possible in hyperspace. But with this, as with the entire fabric of hyperspace speculations, dependence is placed almost entirely upon analogous and symbolic conceptions for evidence as to the consistency and rationality of the conclusions arrived at.

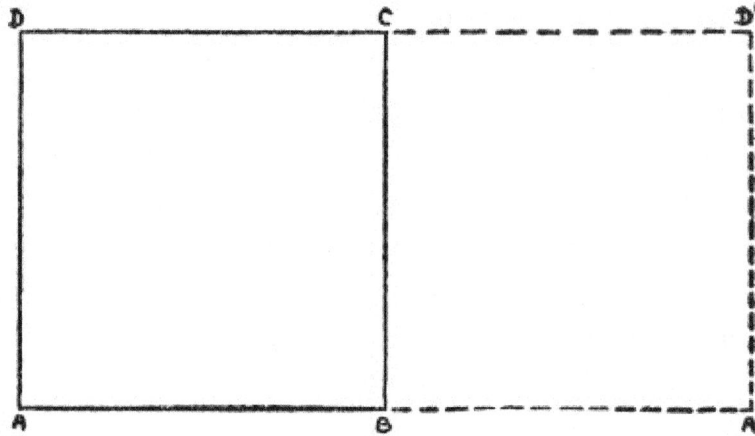

FIG. 13.

It is urged that inasmuch as the rotation about the line *bc* in Figure 13 would be incomprehensible or unimaginable to a plane being for the reason that such a rotation involves a movement of the plane into the third dimension, a dimension of which the plane being has no knowledge, in like manner rotation about a plane is also unimaginable or incomprehensible to a tridim or a three dimensional being. It is shown, however, that the plane being, by making use of the possibilities of an "assumed" tridimension, could arrive at a rational explanation of line rotation.

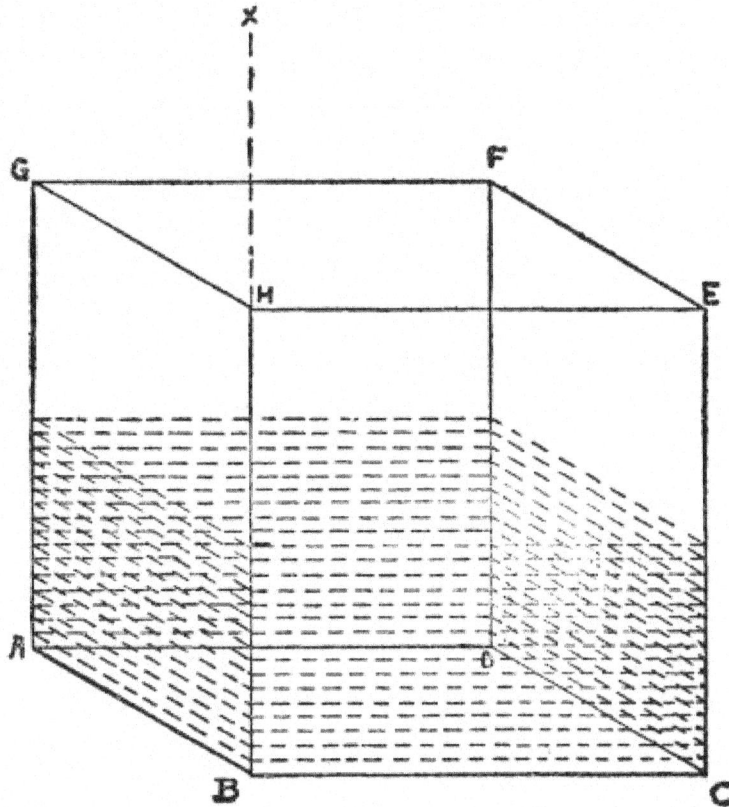

FIG. 14.

Figure 14 offers an illustration by means of which a two dimensional mathematician could demonstrate the possibility of line rotation. He is already acquainted with rotation about a point; for it is the only possible rotation that is observable in his two dimensional world. By conceiving of a line as an infinity or succession of points extending in the same direction; imagining the movement of his plane in the direction of the third dimension thereby generating a cube and at the same time assuming that the lines thus generated were merely successions of points extending in the same direction, he could demonstrate that the entire cube Figure 14, could be rotated about the line *BHX* used as an axis. For upon this hypothesis it would be arguable that a cube is a succession of planes piled one upon the other and limited only by the length of the cube which would be extending in the, to him, unknown direction of the third dimension. He could very logically conclude that as a plane can rotate about a point, a succession of planes constituting a tridimensional cube, could also be conceived as rotating about a line which would be a succession of points under the condition of the hypothesis. His demonstration, therefore, that the cube,

Figure 14, can be made to rotate around the line *BHX* would be thoroughly rational. He could thus prove line-rotation without even being able to actualize in his experience such a rotation.

Analogously, it is sought by metageometricians to prove in like manner the possibility of rotation about a plane. Thus in Figure 16 is shown a cube which has been rotated about one of its faces and changed from its initial position to the position it would occupy when the rotation had been completed or its final position attained.

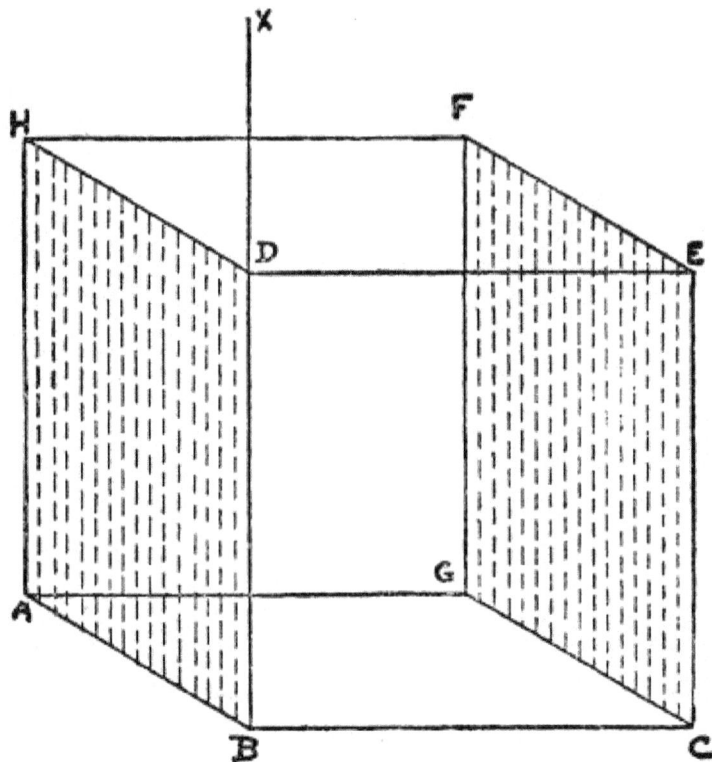

FIG. 15.

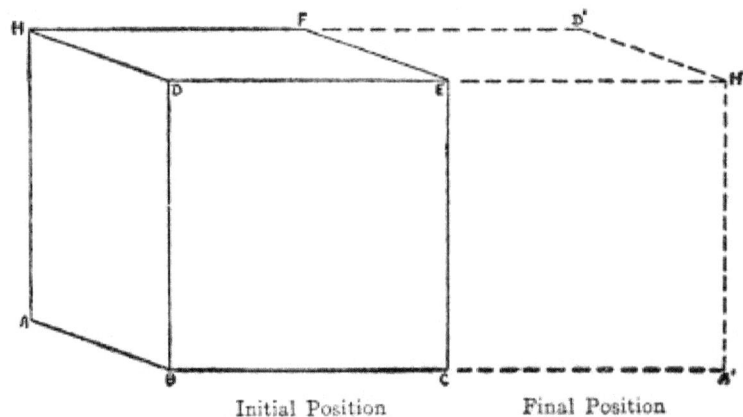

Initial Position   Final Position

Fig. 16.

The gist of the arguments put forward as a basis for plane-rotation is briefly stated thus: The face *cefg* is conceived as consisting of an infinity of lines. A cube, as in Figure 15, is imagined or assumed to be sected into an infinity of such lines, each line being the terminus of one of the planes which make up the cube. Each one of the constituting planes is thought of as rotating about its line-boundary which intersects the side of the cube. The process is continued indefinitely until the entire series of planes is rotated, one by one, around the series of lines which constitute the axial plane. Hence, in order that the cube, Figure 16, may change from its initial position to its final position each one of the infinitesimal planes of which the cube is assumed to be composed must be made to rotate about each one of the infinitesimal lines of which the plane used as an axis is composed. In this way, it is shown that the entire cube has been made to rotate about its face, *cefg*. This concisely, is the "QUOD ERAT DEMONSTRANDUM" of the metageometrician who sets out to prove rotation about a plane. Thus it is made to appear that in order that tridimensional beings may be enabled to conceive of four-space rotation, as in Figures 15 and 16, in which the rotation must also be thought of as taking place in the direction of the fourth dimension, they must adopt the same tactics that a two dimensional being would use to understand some of the possibilities of the tridimensional world.

It is, of course, unwise to assume that because a thing can be shown to be possible by analogical reasoning its actuality is thereby established. This consideration cannot be too emphatically insisted upon; for many have been led into the error by relying too confidentially upon results based upon this line of argumentation. There is a vast difference between mentally doing what may be assumed to be possible, the hypothetical, and the doing of what is actually possible, the practical.

In the first place, plane-rotation in the actual universe is a structural impossibility. The very nature and constitution of material bodies will not admit of such contortion as that required by the rotation of a body, say a cube, about one of its faces. Let us examine some of the results of plane rotation. 1. The rotation must take place in the direction of the fourth dimension. Now, as it is utterly impossible for any one, whether layman or metageometrician, even to imagine or conceive, in any way that is practical, the direction of the fourth dimension it is also impossible for one to move or rotate a plane, surface, line or any other body in that direction. We are in the very beginning of the process of plane-rotation so-called confronted with a physical impossibility. 2. Plane rotation necessarily involves the orbital diversion of every particle in the cube. This alone is sufficient to prohibit such a rotation; for it is obvious that the moment a particle or any series of particles is diverted from its established orbital path disruption of that portion of the cube must necessarily follow. This upon the assumption that the particles of matter are in motion and revolving in their corpuscular orbits. 3. Plane-rotation necessitates a radical change in the absolute motion of each individual particle, electron, atom or molecule of matter in the cube and a consequent retardation or acceleration of this motion. This upon the hypothesis that the particles of matter are vibrating at the rate of absolute motion. 4. It presupposes a reconstitution of each atom, molecule or particle in the cube, changing the path of intra-corpuscular rotation either from a right to left direction or from a left to right direction, as the case may be. The particles of matter in the cube will be acted upon in much the same manner as the particles in a glove when it is maneuvered in the fourth dimension. In describing this phenomenon, MANNING says:[20]

"Every part by itself, in its own place is turned over with only a slight possible stretching and slight changing of positions of the different particles of matter which go to make up the glove."

The slight stretching and slight changing of the positions of the particles referred to would be of small consequence if applied to ponderable bodies. But when used in connection with particles of matter which are themselves of very infinitesimal size means far more—enough, as we have said, to militate severely against the integrity of the cube. It is not deemed necessary to go further into the physical aspects of plane-rotation as it is believed sufficient has been said to negative the assumption from a purely structural viewpoint.

Among the vagaries of hyperspace publicists none is perhaps more notable than the view taken by C. H. HINTON:[21]

"If it could be shown that the electric current in the negative direction were exactly alike the electric current in the positive direction, except for a reversal of the components of the motion in three dimensional space, then the dissimilarity of the discharge from the positive and negative poles would be an indication of the one-sidedness of our space. The only cause of difference in the two discharges would be due to a component in the fourth dimension, which directed in one direction transverse to our space, met with a different resistance to that which it met when directed in the opposite direction."

To be sure. And with equal certainty it might be said that if the moon were made of green cheese it might well be the ambition of the world's chefs to be able at some time to flavor macaroni with it, thus serving a rare dish. Even so, if there were an actual, objective fourth dimension to our space we might be able to shove into it all the perplexing problems of life and let it solve them for us. But the fact that the fourth dimensional hypothesis is itself a mere supposition seems to have been overlooked or rather completely ignored by HINTON. Or else, ought it not be an obvious folly to hope to construct a rational explanation of perplexing physical conditions upon the basis of a purely suppositionary, and therefore unproven, hypothesis?

The recognized domain of the four-space, mathematically considered, is according to the most generous allowance very small, so small, in fact, that the disposition of some to crowd into it the essential content of the manifested universe is a matter of profound amazement. Then, too, it cannot be denied that there is no appreciable urgency or necessity for having recourse to a purely hypothetical construction for explicatory data regarding a phenomenon which has not been shown to be without the scope of ordinary scientific methods of procedure to unravel.

The claim of certain spiritualists, notably ZOLLNER of Leipsig, that the phenomena of spiritism is accountable for on the grounds that the fourth dimension affords a residential area for discarnate beings whence spiritistic forayers may impose their presence upon unprotected three dimensional beings is no less fatuous than the original supposition itself. For upon this latter is built the entire fabric of meaningless speculations so gleefully indulged in by those who glibly proclaim the reality of the four-space. Indeed, clearer second thought will reveal that, when the pendulum of erratic thinking and trafficking in mental constructions swings back, hyperspaces, after all, are but the *ignes fatuii* of mathetic obscurantism.

Then, why should it be deemed necessary to discover some more mysterious realm of four dimensional proportions in which the spirits of the dead may find a habitation? Are the spiritualists, too, reduced to the necessity of further mystifying their already adequately mysterious phenomena? If there were not quite enough of physicality upon the basis of which all the antics of these entities can be explained, and that satisfactorily, one would, as a matter of course, be inclined to lend some credence to these claims; but as it is clear that all organized beings have some power, if no more than that which maintains their organization, and as it ought also be an acceptable fact that such a being is directed by mind; and further, that owing to the nature of a spirit body it can penetrate solid matter or matter of any other degree of density below the coefficient of spirit matter, it ought likewise be unnecessary to go without the province of strictly tridimensional mechanics for an explanation of spiritistic phenomena.

Equally unnecessary and uncalled for is the attempt of certain others who lean toward the view of speculative chemists to account for the none too securely established hypothesis that eight different alcohols, each having the formula $C_5H_{12}O$ may be produced without variation. This is said to be due to the fact that certain of the component atoms, notably the carbon atoms, take a fourth dimensional position in the compound and thus produce the unusual spectacle of eight alcohols from one formula. Have chemists actually exhausted all purely physical means of reaching an understanding of the carbon compounds and are therefore compelled to resort to questionable means in order to make additional progress in their field? It is incredible. Hence the more facetious appears the mathematical extravaganza in which originates the tendence among the more sanguine advocates to make of the fourth dimension a sort of "jack of all trades," a veritable "Aladdin's lamp" wherewith all kosmic profundities may be illuminated and made plain. Not until the perfection of instruments of precision has been reached, and not until human ingenuity has been exhausted in its efforts to produce more refined methods of research should it be permissible even to venture into untried and more or less debatable fields in search of a relief which after all is unobtainable.

Notwithstanding the fact that all attempts at accounting for physical phenomena on the basis of *n*-dimensionality (which is itself by all the standards of objective reference a non-existent quantity and therefore irreconcilable with perceptual space requirements) are to be characterized simply as a senseless dalliance with otherwise deeply profound questions, many have fallen into a complete forgetfulness of the logical barriers inhering in and hedging about the query and have committed other and less excusable errors in the premises. Take, for instance, the suggestion that the action of a tartrate upon a beam of polarized light is due to the assumption of a fourth dimensional direction by some component in the acid. This for the reason that experimentation has shown that tartaric acid, in one form, will turn the plane of polarized light to the right while in another form will turn it to the left. It is not believed, however, that there is any warrant for such an assumption. There is also another kind of tartrate which seems to be neutral in that it has no effect whatever upon the beam of light, turning it neither to the right nor to the left nor having other visible or determinable effect upon it. Indeed, it is not clear how it is hoped to prove such a case by constituting as a norm a hypothesis which is essentially indemonstrable. A more logical procedure would be first to establish the objective, discoverable posture of four-space; show the actual movement of matter and entities therein; locate it by empirical methods of research, and then, basing our assertions upon apodeictic evidences, assume a new attitude toward these phenomena because of the support found in established and verifiable facts. Some hope of gaining a respectful hearing might then be entertained; but at least to do so now appears to be quite untimely.

MAJOR WILMOT E. ELLIS, Coast Artillery Corps, United States Army, in *The Fourth Dimension Simply Explained*,[22] remarks:

"... in the ether, if anywhere, we should expect to find some fourth dimensional characteristics. Gravitation, electricity, magnetism and light are known to be due to stresses in, or motions of, the infinitesimal particles of the ether. The real nature of these phenomena has never been fully explained by three dimensional mathematical analysis. Indeed, the unexplained residuum would seem to indicate that so far we have merely been considering the three dimensional aspects of four dimensional processes. As one illustration of many, it has been shown both mathematically and experimentally that no more than five corpuscles may have an independent grouping in an atom."

The weakness of this view may be due to the fact that at that time MAJOR ELLIS was emphasizing in his own mind the necessity of simplifying the conception so as to make it of easy comprehension rather than the establishment of any fealty to truth or the spirit of mathesis in his examination of the problem. What therefore of reality the student fails to find in his view may be attributed to the sacrifice which the writer (MAJOR ELLIS) felt himself called upon to make for the sake of simplicity. Hence a certain expressed connivance at his position is allowable. But, on the other hand, if such were not the conscious intent of MAJOR ELLIS it is not understood how it should appear that "the unexplained residuum would seem to indicate that so far we have merely been considering the three dimensional aspects of four dimensional processes." Contrarily, it has yet to be proved that three dimensional space does not afford ample scope of motility for all observable or recognizable physical processes and that there is no necessity for reference to hyperspace phenomena for an explanation of the "unexplained residuum." It is, of course, understood that many of the possibilities predicated for hyperspace are purely nonsensical so far as their actual realization is concerned. Our concern is, therefore, not with that class of predicates, but with those wherein reside some slight show of probability of their response to the conditions of n-dimensionality either as a system of space-measurement or a so-called space or series of spaces.

MAJOR ELLIS concludes his simple study of four-space by proposing the following query:

"May not birth be an unfolding through the ether into the symmetrical life-cell, and death, the reverse process of a folding-up into four dimensional unity?"

It is confessed that there seems to be nothing to warrant the giving of an affirmative reply to this query. It is, perhaps, sentimentally speaking a very beautiful thing to contemplate death as a painless, unconscious involvement into a glorious *one-ness* with all life, and birth, as the reverse of all this. But where is the utility of such a dream if it be merely a dream and impossible of realization?

SIMON NEWCOMB,[23] at one time one of the outstanding figures in the early development of the fourth dimensional hypothesis, openly declared that "there is no proof that the molecule may not vibrate in a fourth dimension. There are facts which seem to indicate at least the possibility of molecular motion or change of some sort not expressible in terms of time and the three coördinates in space."

Of course, there is no proof that a molecule may not at times be ensconced in a four-space neither is there proof nor probability that it is so hidden. Indeed, there is no proof that there is such a thing as a molecule for that matter.

In all of the foregoing proposals it is assumed that the fourth dimension really exists and that it lies just beneath the surface of the visible, palpable limits of the material universe; that lying in close juxtaposition to all that we are able to see, to hear or sense in any way is this mysterious, eternally prolific, all-powerful something, hyperspace, ever-ready to nourish and sustain the forms which have the nether parts firmly encysted in one or the other of her *n*-dimensional berths. Thus it would seem that while yet functioning in a strictly tridimensional atmosphere, some one, more reckless than the rest, should at last stumble upon some up-lying portion of it and be instantly transformed into a mathetic fay of etherealized four-dimensional stuff.

PART TWO

SPATIALITY

AN INQUIRY INTO THE ESSENTIAL NATURE OF SPACE AS DISTINGUISHED FROM THE MATHEMATICAL INTERPRETATION

CHAPTER VI

Realism Is Determined by Awareness—Succession of Degrees of Realism—Sufficiency of Tridimensionality—The Insufficiency of Self-Consistency as a Norm of Truth—General Forward Movement in the Evolution of Consciousness Implied in the Hyperspace Concept—The Hypothetical Nature of Our Knowledge—Hyperspace the Symbol of a More Extensive Realm of Awareness—Variations in the Method of Interpreting Intellectual Notions—The Tuitional and the Intuitional Faculties—The Illusionary Character of the Phenomenal—Consciousness and the Degrees of Realism.

Things have value for us only to the extent to which we can become aware of their being. The appraisement of all objects, conditions, states or qualities is determined directly by the degree or quality of awareness with which we apprehend them. Those elements which are without the intellect's scope of awareness have no interest and hence no value so far as the individual intellect is concerned. And this is true of all degrees and states of consciousness from the lowest to the highest, from the human to the divine.

There enter into all conscious determinations three factors, namely: (*a*) the scope, or totality, of adaptations which an organism can make in the sensible world, (*b*) the power of consciousness to make adaptations and (*c*) environment. These three are interdependent. The totality of adaptations depends primarily, of course, upon the quality of conscious powers or faculties, and also, in a lesser degree, upon opportunities afforded by environment. Faculties of consciousness are derived directly from the influences exerted upon the organism by his environment and the results of the struggle to overcome them. Environment is of two kinds, artificial and natural. The artificial environment is such as has been modified by our conscious action upon external phenomena. The residue is natural. And thus the scope of adaptability becomes an unvarying witness to the quality of consciousness manifesting through a given organism.

The universe is so constructed that the essential character of its various states and qualities is a fixed quantity for a given scope of consciousness and varies only as the sphere of consciousness varies. States of existence or scopes of adaptation which are registering upon a higher plane or in a more subtle sphere of existence than that in which we may at any time be functioning can only appear evidential to us when the mechanism of our consciousness becomes congruently adjusted therewith. So that the focus of consciousness must always be a variable quantity adaptable, under proper conditions, to any plane in the kosmos. Consciousness, then, becomes the sphere of limits both of knowledge and adaptability. But lest we seem to admit implicitly part of the contentions which mathematical publicists have made in postulating the unodim and duodim consciousness, it is necessary carefully to differentiate between the results arrived at as a result of the two procedures. In the first place, analysts *assume* the existence of a unodim and duodim plane of consciousness and proceed to construct thereon an analogy designed to show the feasibility of another assumption, the fourth dimension. While, in laying the foundation of consciousness upon a tridimensional plane we do not start with an *assumption*, but with a fact. Therein lies the difference. Enormous advantages inhere in a procedure based upon facts, but in a series of planes built upon assumptions no such advantages are discovered. For however much the series of hypothetical planes may be extended or elaborated there must inhere necessarily throughout the series an assumptional value which vitiates the conclusions no less than the premises. The sanity and integrity of intellectual operations depend almost entirely upon the differentiation which we make between the necessities arising out of assumptions and those which spring up empirically from established facts. No procedure is necessary to establish the value of such a differentiation, nevertheless it may be suggested that it is allowable, under the rules of logic, to make any assumption whatsoever so long as care is taken to see that the conclusions embody in themselves the characteristics of the original premise. For instance, it is permissible to assume that space is curved. Under such an assumption, it is only necessary that the constructions which follow shall be self-consistent. But the case is different when we come to deal with spatiality and vitality. These are quantities which cannot, in the last analysis, be made to conform to the rules of the game of logic.

Thus, when it is intimated that realism lends itself to an apparent division into degrees, and that each degree has a corresponding state of consciousness, it is by no means to be inferred that such apparent divisions are of mathematical import. For, in reality, i.e., when the consciousness has expanded so as to become congruent with the limits of even the space mind (vide Fig. 20), there appear to be no divisions in realism. It is only because of the fragmentariness of our outlook upon the kosmos that realism appears to be divided into various planes; for all of these planes are one. The divisions exist for relative knowledge, but not for complete knowledge; they exist for a finite intelligence, but not for a transfinite intelligence. That is why we view realism as a series of planes. It is because we discover that, as we proceed, as our consciousness expands and we take in more and more of the vital activities of the kosmos and understand better the causes underlying that which we contact, we have passed from a state of lesser knowledge to one of greater knowledge. And so we say we have passed from one degree of realism to another, whereas, really we have not passed from one degree of realism to another degree. Instead, it is our consciousness that has expanded.

If now, we conceive reality to be a scale extending from one extremity to another (that is, from supreme consciousness to entire unconsciousness, from final knowledge to total ignorance), and the intellectual consciousness as the indicator which traverses the scale denoting at all times the precise degree of our comprehension of reality, and hence the degree of expansion of consciousness, we shall constitute a similitude closely approximating the real *status quo* of humanity with respect to the sensible and supersensible worlds. The quantity or force which causes the indicator to move along the scale is the quality of awareness. And this varies directly as the scope of adaptability varies. Realism is homogeneous throughout its extent; but the scale marked upon it registers from *naught* to *unity*. And between these every conceivable degree of awareness may be registered. The indicator moves only as the scope widens, and thus is shown a change in the quality of awareness. For, however paradoxical it may seem, the wider the scope of knowledge the better its quality: the more one knows, the more complete and of higher quality becomes that which he knows.

The intellect is of scientific tendence, studiously rejecting all phenomena which do not yield to its senso-mechanisms. Even intuitions suffer the humility of rejection and do not escape the limitations which the intellect imposes upon them. This is so, because, as yet, there is no adequate perceptive and conceptive apparatus for the propagation and classification of intuitions, as apart from concepts. The outcome of these proscriptions is that intuitions—free, mobile, and more or less formless in themselves, must first be rehabilitated and vestured in garments *a la intellect* to conform to the prevailing mode. But intuitions thus treated are no longer intuitions, but empirical concepts. True intuitions are like aqueous vapor—amorphous, permeating, diffusive: axioms or empirical concepts are like cakes of ice—formal, inflexible and conforming to the shape of the mold into which they are poured. Because of this—the scientific tendence of the intellect and the consequent necessity of reforming so much of the data which constitute its substructure, of pressing, condensing and reshaping it to suit its own ready-made patterns—it can be perceived how profound is the influence of the intellectual consciousness in determining the character of the totality of data which the sensible world, and for that matter, the supersensible, offer us. The intellect is the only means at hand for the interpretation of the meaning and significance of the world of phenomena. Consequently, whatever meaning or significance we are led to attach to that part of the universe which we contact, in any way, is dictated by the intellectual consciousness. There is no escape from the decisions of the intellect so long as the present scheme of things endures.

Thus, by whatever standard of reference the matter may be determined, it remains indisputably established that the intellectual consciousness is the sole determinant of the phenomenal value of everything within our scope of awareness or adaptability. And whatever the fault, incongruity or discrepancy that may be revealed by a more intimate knowledge of the genesis and character of the appearance of the sensible world, it will be found to be due to the peculiar cut and mode of the intellect and not to things themselves. The value, qualitative or existential, which the intellect irrevocably assigns to objects and conditions in the world of the senses is the exclusive *norm* not only by which these are judged, but also, by which our action upon them and their action upon us are determined. Images or objects which do not act upon us and upon which we cannot act have no interest for us. But as an integral part of the totality of images or objects in the sensible world, we must inevitably act upon all that is outside of ourselves, and these, in turn, must react upon us. On the other hand, there must be objects and images in the universe of life and form upon which, because of their inherent nature and on account of the lack of our interest in them and their interest in us, we can neither act nor become the object of their action.

But herein is a mystery. For, either we act upon and are recipients of the action of the totality of images or objects in both the sensible and supersensible worlds, or we are so placed in the grand scheme of things that both ourselves and the sphere of our interests and possible actions are closed circuits, hermetically sealed and non-communicative with the other like spheres, which do not and cannot act upon us. There is yet a third possibility—that we are so fashioned, in the entirety of our being, that some part of us is exactly congruent with some part of every sphere of possible actions and interests in the kosmos, and therefore, each of us has being or consciousness of a kind that is keyed to and registering in the totality of such spheres; and that, at present, because our interests and possible actions are limited to the plane of sensibility, we are conscious only there. And further, that although those spheres of our consciousness which are fixed to register in other planes do not answer to the lowest on which we now operate, having a character of which we are unaware, they nevertheless cannot be said not to exist, because of the lack of communication between them. Among these three possible choices, we have no hesitancy in expressing a decided preference for the last mentioned—that the range of our being is co-extensive with the range of reality, and like a pendulum, we oscillate, at long intervals, between two kosmic extremities—nescience and omniscience.

The intellectual consciousness is the touch-stone of realism. It is like a spreading light which, as it expands, reveals previously darkened corners and conditions, only it has power both to reveal and to bring into manifestation. In its present state, man's consciousness is like a searchlight. It illumines and takes cognizance of everything that falls within its scope of motility and is consequently able to study in detail that which it reveals. But that which is beyond its scope is as if it never existed so far as the individual consciousness is concerned. It is not reasonable to predict that the same characteristics that are observable in any given state shall persist throughout all the various scopes through which the consciousness must proceed in its evolutionary expansion. For the scale of kosmic realism is one grand panorama extending from the grossest to the most subtle and refined. While in general the thread of realism may pervade the entire scale it is nevertheless marked by many and diverse changes in its characteristics as it is followed from one stage to another. So that the realistic character of one stage may vary greatly from that which next preceded it or from that which will succeed it. It would appear, therefore, that in passing from one stage of realism to another there need not remain anything but the mere fact of reality in its connection with ultimate reality; for it is obvious that in every condition of realism which may be encountered in the kosmos there must be a basic thread of ultimate reality running through the whole. The entire gamut of realism may accordingly be traversed without the danger of being diverted from the golden thread of realism which thus permeates all. It is always the phenomena of realism with which we are concerned and which we are trying to understand rather than realism itself. It is this that confounds us. If it were not for the phenomena, which is the way realism or life presents itself to our consciousness, we should experience no trouble in discovering the reality, all other things being equal. For the former ever obscures the latter. It is the supreme task of mental evolution to break through the clouds of phenomena in the search for the eternal substratum of reality which runs through the sensible universe of things.

The first view of conditions that the mind takes upon awakening to consciousness in any new sphere of cognition is necessarily hazy and inchoate. There is more or less of astonishment, wonder and bewilderment upon first becoming aware of a new scope of realism. In this state it is natural that the mind should overlook or ignore much that is essential and perhaps all that is so even escaping the true import of the phenomena which it senses. It is reasonable, too, that in such a state the main outlines of what is really seen may be greatly distorted and exaggerated so that it is well-nigh impossible to secure a correct comprehension of the character of a new scope of realism from any early survey. It is not until later years, after much study and circumspection that the mind, becoming used to the new conditions, begins to get correct impressions and to make valid judgments as to that which it discerns. And even then, it not infrequently happens that the resultant view of things in general is found to be in need of revision and correction. Hence, after everything is sifted down to the ultimate allowance for the illusion incident to too much enthusiasm and wonder we have only a very small residuum of truth upon which to build and this latter we often find to be the single thread of reality which runs through all the phenomena and which is, therefore, the only quantity that remains worthy of much consideration.

Thus it is with religion. The path of progress over which our religious conceptions have come need not be outlined here, but to any one at all acquainted with the history of religious thought and ideals it at once must be patent that it has been one continuous surrender of the old for the new, of one degree of realism for another newer and higher degree; that always it has been the phenomena, the flora of the ideals which have had to give way, while nothing was left but the roots of realism from which they have sprung. It has been the same with scientific knowledge. Facts have been collected and hypotheses proposed to synthesize them and yet these have had to give way for others, and others still, until the data of scientific knowledge to-day are quite different from what they were in earlier days. And yet permeating the scientific knowledge of all times has been the golden thread of reality, and of all facts and systems of facts which man has successively assumed and surrendered nothing has remained but the reality; indeed, nothing could so remain, but reality. So it is with air phenomena with which consciousness has to deal. This perhaps is due to the fact that the mind interprets phenomena in accordance with the quality of its awareness, and as consciousness is a variable quantity, its standards of interpretation will likewise vary. Each new scope of awareness, after this manner, yields higher and more exact standards of interpretation. And then, progressing in awareness from the segment to the whole a fuller view of the phenomena as well as of reality itself is gained and also a more comprehensive judgment of the relations which exist between the segment and the whole. In other words, as the scope of consciousness widens it becomes more and more apparent that what was first thought to be a separate segment is in reality identified with the whole in an indissoluble manner. For the Thinker is then not only aware of the segment as such, but he is also conscious of the fact that it has definite relations with the entirety and that what he needs is merely a more extended consciousness.

In denying the existence of the four-space or spaces of *n*-dimensionality as described and defined by geometricians, we do not thereby deny the existence of a plane of consciousness which is as much unlike the conditions of the tridimensional world as it is said to be unlike the four-dimensional world; but what we do deny is that such a higher plane of existence has necessarily to be conditioned by such characteristics as the metageometricians have proposed. It is maintained that there is no basis in consciousness for a world of four dimensions; that the consciousness has no tendency for action in four-space. Neither has matter nor life any inclination or potency to behave in a four-dimensional manner. It is indeed more rational to suppose that there is a higher plane, in fact, a series of higher planes, in which the thread of realism is continuous, not broken as it necessarily would have to be in extending to hyperspace, nor curved as in a manifold; that this series of subtler and finer planes of consciousness are merely an elongation of our three dimensional scope of realism. It, therefore, remains only to master the phenomena of each in just the same manner as we have, in a measure, mastered the phenomena of tridimensionality. For it is easily conceivable that the quality of consciousness is such that it may adapt itself to a far wider range of possibilities than may be discovered in hyperspace and still be a tri-space quantity.

It is believed, however, that in all the new and higher planes of consciousness tridimensionality is the norm both of the phenomena and of the reality peculiar to them. And that, being such, does not change or vary, but is a fixed quantity regardless of the plane of consciousness. Furthermore, it is believed that the highest state of consciousness in the entire kosmos could easily exist, and does so exist, upon the basis of three-space as the norm of its extent.

A sharp line of demarkation should be drawn between the reality which is life and consciousness and that which belongs to the realm of phantasy. For it is the prerogative of the intellect to create, out of the remains and deposits which it finds in the pathway of life, whatsoever it wills. This it does continuously; but it scarcely can be expected that such creations shall be endowed with the same dynamic character as that which life bestows upon its creations. The creations of the one are merely dead carcasses while those of the other are vital and real. Between them the same marked difference exists as between the growing tree and the lumber which the builder converts into a house. The organization which we witness when we look upon a building made of the dead body of a tree is not the same kind of organization as that which we see when we view the living, growing, vital tree. The dead tree is a deposit of life cast off by it when it passed on. Whatever the intellect can do in disposing of the remains of the tree-life is conventional and artificial. If it convert it into an edifice it will then bestow upon it a sort of consistency which is quite sufficient for all purposes. But the consistency which holds the organization of an edifice together is not the kind of consistency which holds a living tree together. In fact, there is a consistency that is not consistent. Such is the consistency of metageometry. It is self-consistent and yet inconsistent with the consistency of the kosmos and its norm of being which is consciousness.

Self-consistency is one thing and kosmic consistency is quite another. It does not necessarily follow that because a given scheme of thought is consistent in all its parts that it is also consistent with universal truth or with life. This very vital fact was overlooked by GAUSS and all those who followed in his wake when he discovered that his *Astral Geometry* was consistent throughout in all its parts. There is only one norm of truth and that is kosmic consistency. It matters little that a thing shall be self-consistent; it matters much whether it is consistent with the universal standard. It has been shown to be logically possible to elaborate at least two different systems of geometry, namely, the geometry of the acute angle and that of the obtuse, which, while each of them is self-consistent throughout, are nevertheless inconsistent with each other and with the geometry of the right angle (Euclidean). This, it would seem, appears to be sufficient for the invalidation of either one or both of the non-Euclidean systems of geometric thought. Indeed, if it can be shown that the Euclidean geometry is more representative of the true approach to the norm of space-genesis and of creation so far as its mode of manifestation is concerned, and consequently true of the norm set up by consciousness, the rejection of both systems of non-Euclidean geometry seems to be thoroughly warranted. But this is obvious and requires no demonstration nor comment to make it clear.

We have only to ask ourselves whether it has ever occurred to us that consciousness has either a tendency to or adaptability for action in a curvilinear manner; or, if when we contemplate ideas or idea-relations we have the impression of perceiving a curvilinear or manifold tendence in them either of a positive or negative nature, and also whether it has been observed that our thought processes naturally assume four-dimensional attitudes. If we find that such a query must be answered negatively, and indeed we must so find, then, we have no basis for the assumption that any one of the systems of non-Euclidean geometry is valid either for the present status of consciousness or for a future existence, since it is true that the future is but an elongation of the present. Evolution is to bring no radical changes in the norms of reality; it has merely to deepen and widen and make more intense, efficient and comprehensive the present scope of our consciousness and thereby, while the Thinker is passing from one degree of realism to another, to bring him into a clearer conception of what his own limited scope of motility means to the whole.

The four-space is a mathetic divertisement. That is, it cannot be said to lie in the direction of a straight line which proceeds either in a forward or lateral direction. Neither does it lie in a plane which is vertical or horizontal to the sensorium. It is, therefore, a fractural departure from any conceivable tridimensional direction, a geometric anomaly. Evolution, despite the minor aspects of its movement, undoubtedly proceeds in a straight line and not by a zigzag nor discontinuous line and hence it is irrational to assume that it will, when it passes on to the next advanced stage, emerge into the realm of the four-space. For the so-called hyperspace of geometry cannot by any standards of reference be said to lie in the plane of any straight line which can be described in three-space. If life is to evolve more efficient forms and if the forms are to evolve into more perfect organizations and mind and consciousness to become more intense and comprehensive expressions of the divine mind of the kosmos it is certainly not in the domain of hyperspace that these shall find the substructure of their higher development; but, if at all, it shall be found, as in all times past, in the realm of perceptual space where bodies are said to have three and only three possibilities of motion.

What then is the significance of the more than a thousand years of mathematical labors; of all that has been said and done in an endeavor to bring into the popular consciousness a conception of hyperspace? Is it a question of *"Love's Labour's Lost?"* Or is it a mere prostitution of mathematical talent? To answer these queries is the burden of this treatise and it is hoped that as the text continues the reader may be able to arrive at his own conclusions as to the relative value of the work of the mathematicians in this respect and be able to judge for himself the true significance of it all.

The specific value of consciousness as a determinative factor in space-measurement has been recognized by all who have sought to arrive at a logical justification for the conception of four-dimensionality by analogous reasoning. The existence of the *unodim* with consciousness limited to a line or point has been assumed and it has been shown how greatly such a being would be handicapped by his limited area of consciousness, it having been proposed to confine his consciousness to one dimension. An *unodim* would, of course, be entirely unaware of any other dimension than that in which he could consciously function. So that with respect to his own consciousness no other dimension would be necessary for the continuance of his life processes. He might live his life without any knowledge even of any limitations or barriers to other things higher than those of his plane. He would be content to exist in the one-space and enjoy the benefits which it offered. He could have no notion of the two-space, but it has been allowed that a *super-unodim*, an *unodim* metageometrician, if you please, could reason out a mental conception of what the two-space might be. Passing on to a space of two dimensions, the domain of the *duodim*, a greater freedom of movement is allowed and instead of being able to function in only one dimension the inhabitants of this plane would find themselves able to move about in at least two directions. Consciousness would accordingly enjoy a more comprehensive scope. But in a manner similar to that used by the *unodim* metageometrician it is held that the *duodim* could get a conception of the three-space by analogous reasoning and so gradually become conscious of a higher degree of spatiality than his own. In the conscious reasoning of both, however, is the condition of perpendicularity. That is, it must be assumed by both the *unodim* and the *duodim* that the new dimension must lie in a plane perpendicular to their space. So, the *unodim* would postulate that the two-space must lie in a direction at right angles to his space, and yet he would not be able to indicate the direction owing to his ignorance of any experience that would acquaint him with the new space as well as the want of possibility of motion therein. Similarly, the *duodim* would arrive at a conception of three-space. Thus, it has been argued that *tridims,* or people living in our tridimensional world, could, by using a like line of argument or reasoning, arrive at a conception or understanding of the four-space, which, of course, must also lie in a direction at right angles to three-space.

The implications of this mode of thought show how thoroughly the mathematician recognizes the limitations which consciousness imposes upon our knowledge of the world and the subtler conditions about us. But, moreover, it is even obvious to all who stop to think about it; for it can readily be seen how little those things which do not enter our scope of awareness affect us either physically, mentally or spiritually. But no one can be so bold as to deny utterly that anything exists but what is found in our consciousnesses. It is even true that in the great centers of population where people are compelled to live many families in the same house, it is the usual thing for these individual families to live in complete forgetfulness of all the others in the house and live their lives so completely that it would be exceedingly difficult to measure the effect the one has upon the other. The mathematician, as is shown by the hyperspace movement, recognizes that there are planes of superconsciousness the nature and character of which persons confined to limited areas of consciousness can have no knowledge and may only arrive at that knowledge by serious thought and contemplation. In other words, they tacitly admit the existence of higher planes of consciousness as well as the necessity of elevating the personal consciousness in order to comprehend them. Although it was not expressly allowable in the analogy of the *unodim*, it is nevertheless one of the strongest implications of the process of reasoning that the *unodim* could have easily raised the plane of his consciousness by continuing his researches until he, too, became conscious of the three-space, mathematically, as well as the two-space. For it was not necessary for him to raise the plane of consciousness in order to contact the two-space. He had need only to widen it. But in order to comprehend the mathematical three-space it would be necessary for him to elevate his consciousness.

The fundamental error in the foregoing line of thought rests in the fact that awareness in the human family has not developed in the manner outlined. The human species has not come into conscious relations with the three-space by outgrowing the one-space and the two-space in succession. The fact of the matter is that when consciousness first dawned it must have encompassed all three dimensions simultaneously and equally and there is nothing to indicate that its rise was otherwise. Then, specifically there is no evidence that the evolution of consciousness has proceeded in a rectangular manner. Indeed, there is undoubtedly no warrant for the assumption that it has progressed in ways that are mathematically determinable at all. The question very naturally rises in view of the above as to the relative value of mathematical knowledge in the scheme of psychogenesis. Can mathematical knowledge or laws be said actually and finally to settle once for all time any question in which consciousness or life enters as a factor? Upon the response to this question hinges unanswerably the decision as to the category which mathematical knowledge should by right occupy in the entire schematism of life. If it can be successfully maintained that final judicative power abides in mathematics in the determination of these questions, then it would be useless to struggle against the fiat of mathematics and mathematicians; verily, we should be compelled to accept *nolens volens* all that mathematicians have devised about hyperspace and its connotations. If, on the other hand, it can be shown that no such judicative power inheres in mathematical knowledge we shall then be able to establish for mathematics a true category and to dispose of the hyperspace movement in a manner that shall at once be logical and necessary.

That the discovery of hyperspace by the mathematician is merely an aspect of a general forward movement in the evolution of consciousness can be shown by a brief correlative study. Hyperspace is the artificial symbol of a higher and more extensive realm of awareness. For it cannot be doubted that to be able to think in the terms of hyperspace, to study the various relations and interrelations upspringing from the original premises, actually to become conscious in the hyperspatial realm thus constructed, requires a different species or quality of consciousness than that required for ordinary thinking. The period covering the rise of artificial spatiality is contemporaneous with the rise of the phenomena identified with the spiritual life of SWEDENBORG; for during the same time he began a series of visions which revealed to him great knowledge of the unseen and supersensuous realities of life and existence. His consciousness was being flooded with the light from so-called celestial spheres and he was gradually becoming conscious of a "new dimension," a new space, a higher world that is altogether unlike the world of the senses. During this period, too, DANTE, the great kosmic seer began to jot down the results of his "hyperspace" experiences, after which he wrote his *Divina Commedia* in which he describes more or less minutely some of the characteristics of the hyperspace domain which was revealed to his consciousness as he saw and interpreted it. Both SWEDENBORG and DANTE being deeply religious and pious-minded had their reports of the new world colored by their own mental experiences and proclivities. PLATO had at an earlier day set down what he conceived to be the ethical and civic characteristics of the new age, the *Utopia* of mankind living on a higher plane. It was during these days that the doctrine of evolution was born, although it remained for DARWIN to formulate and buttress it with a stupendous congeries of facts. MARTIN LUTHER, the great religious reformer, likewise contacted the radiating light-glow of a higher consciousness into which the race was coming but of which only the foremost were able to get advance glimpses. KANT, one of the peerless leaders of the vanguard of humanity, at this time also, conceived and wrote his *Critique of Pure Reason* which is likewise an evidence of the upliftment of his consciousness on the side of pure intellectuality and the commencement of a general period of illumination. And then, later, but embraced within the same period, artists began to get glimpses of this higher consciousness which showed itself in a new and strange departure in art. In rapid succession new schools sprang up and came to be known as the "cubist," "post-impressionist," "futurist," "orphimist," the "synchromist" and the "vorticist." Art really began the search for the "plastic essence of the world" trying to portray its conception of the "real image of the spirit" of the world. Color acquired a new kind of splendor and painting gave birth to a new intrinsic beauty of material and sheer loveliness of texture. All of which were evidences of an intellectual up-reaching in response to the sharp appulsions from above. DARWIN's mind, being of scientific bent, saw and interpreted everything in the terms of materialistic science; but there is no doubt but that the expansion of the area of awareness which his mind experienced in his great conception of evolution as a continuous process and all that it implies thereby was a result of the universal appulsion of the human intellect against the new domain of consciousness. And KANT's conception of space in general perhaps may be said to have been the seed-thought for the metageometrician.

But thus it will be noted that in all the cases mentioned in the foregoing there is always present the personal element of the investigator, and that the reports of each of these have been colored and characterized by their individual consciousness and experiences. That all reports would agree with respect to details connected with the new domain of consciousness could scarcely be expected owing to the wonder and bewilderment which seize the intellect under such circumstances. No implication that the mathematicians have been unduly excited by what they have discovered after years of patient research in this direction is indicated by the foregoing observations; but it cannot be denied that the enthusiasm of the moment and the consequent minimization of all other phenomena but the special line being investigated serve very effectively to obscure the mental vision of the more partisan. It perhaps is sufficient that the investigator should set down in as orderly manner as possible the things which he conceives, and that he should interpret them according to the standards of his own intellect. More than this cannot be expected. Moreover, it usually suffices that the future investigator, far removed from the beclouding influences of partisanship, who successfully raises his consciousness to that higher plane shall be able to synthesize the findings of all and thereby with the aid which comes to him from a more advantageous position arrive at sounder views and a more reliable judgment.

It will thus be seen that the metageometrician's method of interpretation is no more entitled to final credence and general acceptance than that of the spiritualist, SWEDENBORG, or the occult seer, DANTE. For in their best moods and at their highest points of mental efficiency these have only succeeded in vaguely symbolizing what they have conceived of the realities of the supersensuous realm in terms of their own experiences. Is there any more cogent reason, then, for accepting the analyst's conception of a world of hyperspace peopled with ensembles, propositions, spaces of *n*-dimensionality and other mathetic contrivances than the *Inferno* of DANTE, inhabited by hideous shapes and repellent denizens, the remains of ill-spent earth-lives or SWEDENBORG's *Celestial Realm*, wherein dwell numerous beings of celestial character performing various tasks in the work of the world? These observations should not lead the reader to come to the conclusion that the visions of DANTE and SWEDENBORG are deemed to be more worthy of credence than mathematical knowledge when that knowledge is limited to the sphere where it rightfully belongs; but the proper view is that which would make it appear that it is the way these widely differing workers interpret what they have seen; that it is the adaptation of the unseen realms to the peculiarities of the mentalities which observe them. The mathematicians have simply portrayed as well as they could their conception of the new stage of consciousness and its contents, and following the *modus vivendi* of all intellects have interpreted these things in the terms of mathematics, merely because mathematics constituted the best available symbology at hand for the purpose. Similarly, the painter sees a new world of color; the politician, a new era of political freedom; the religious enthusiast, a new religious conception; the scientist, a new condition of matter and energy, and so on, to the most ordinary mind, each sees something new while at the same time is necessarily limited to the confines of his own mentality when he comes to interpret what he sees and conceives. Hence, there would appear to be only one way to regard all these advances and that is by synthesizing them, by correlating, and by tracing them to a common source, and finally by seeing them as one general forward movement of intellectual evolution.

Man, the Thinker, who in essence is a pure intelligence, has two mental mechanisms or organs of consciousness. One of these is the brain-consciousness or the egoic. It is so called because the brain is its organ of expression and impression. It manifests through the brain and uses it to further the various objective cognitive processes. The brain-consciousness is a child of the physical body and its life is intimately identified with the life of the body. This consciousness may be called the *a posterioristic* mechanism or organ of the Thinker and is therefore his means of interpreting the phenomena of the objective world. Cell-consciousness is a phase of the ergonic functions of the *a posterioristic* mechanism. The other organ of consciousness is an aspect of the intelligence of the Thinker himself and perhaps may be said to be the active, organized portion of that intelligence. It is separate and distinct from the *a posterioristic* consciousness yet sustaining a substructural relationship with it, being the source of the egoic or brain-consciousness. It may be called the *a priori* consciousness. Its roots are buried deep within the heart of the space-mind and it is therefrom nourished and developed by what it receives in the way of intelligence. It is the intuitional faculty; knows without being taught; conceives without reason; interprets according to the norms of the space-mind or the divine mind of the kosmos. It always resides on a higher plane than that of the brain-mind or consciousness, only at rare intervals being able to contact it with flashes of its own intelligence as intuitions.

The *a priori* consciousness being the intuitive faculty of the Thinker is, therefore, a phase of his mental life on a higher plane than the sensuous. All its conceptions constitute the *a priori* knowledge of the brain mind so-called. The *a priori* faculty of man's higher consciousness gives the character possessed by that form of knowledge known to philosophy as the *a priori*. So that the *a priori* has a more substantial basis than has hitherto been surmised. It is not only that which may be said to transcend experience but that which is the organ of contact with the supersensuous realities as well as of expression through the brain-consciousness.

The mind's method of apprehending objective phenomena is not a direct process but an indirect process by virtue of which neurograms or nerve-impacts registered in the brain are interpreted. External sense-impressions are, of course, conveyed to the cortical area by means of appropriate vibrations which traverse the lines of the neural mechanism. These are recorded in the brain areas just as a telegraphic communication is registered in the apparatus of the receiving end, and in being so, they make terminal registrations which man, the Thinker, interprets after a psychic code which has been built up empirically. That is, he comes to know that certain rates of vibration and certain peculiarities therein mean certain things when referred to the sensorium. He then interprets according to this experience the symbolism of all neurographical impressions. But it is obvious that under such circumstances, where the interpreter is far removed from the thing itself and finds it necessary to interpret rates of vibration or symbols in order to arrive at a knowledge of the intelligence which is conveyed to him, the chances of inadequate conception are very great indeed. In fact, it is not possible through the use of neurographical symbols alone to get any complete notion of the phenomena considered. And thus there stands between the Thinker and absolute knowledge a barrier which prevents his arriving at a state of certitude in his knowledge of the world of sensible objects. It is, however, a barrier which will always remain, checking ever his approach to finality in his understanding of the universum of appearances.

A markedly different condition obtains in the realm of the *a priori* or intuitional for the reason that the barriers which inhere in the neurographical or *a posterioristic* method are absent and the Thinker has a more direct approach to the objects of cognition. Hence the chance for error is very small indeed. This will account, therefore, for the superiority of the intuitional over the rational or the perceptual. Indeed, it is doubtful whether the purely rational possesses any value whatsoever when its *modus vivendi* is unsanctioned by the intuitional.

Else why can we not be certain that the results of our rational processes are correct at all times? Is it not because we lack the power to perceive whether our premises are correct in the first place? Quite truly. For if the Thinker can intuit the necessity and certitude of any given premise it follows that the consequences of that premise are true. It would, therefore, appear that the more the intuitional faculty is developed the clearer will be our perceptions not alone of abstract values but of objective things themselves. Further, it is doubtlessly true that the more the space-mind is developed in the human race the deeper will become our perceptions of the essential *be-ness* of things so that whatever may be the presentations of the space-mind to the brain-mind they will be by far more accurate than the impressions we receive through the latter as a medium of apprehension. It is but natural, however, that in the present more or less chaotic condition in which the faculty of the intuitional is found it should be difficult even to interpret its presentations accurately. It is perhaps due to the fact that we are unused to its deliveries and mode of registration as well as to the fact that it has been overshadowed by the intellectual or rational faculty. But the mere fact that it is present and functioning, even if but rudimentarily, is evidence of its potentiality and the possibility of its future development to a still higher degree of efficiency.

There is no doubt but that the original impulse which resulted in the creation of the faculty of perceptibility in the Thinker also marked out the metes and bounds of our entire range of perceivability which includes not only the intuitional but something higher still. There is no doubting either the obvious fact that these metes and bounds cannot have been other than rudimentary or general lines of denotation, and that the work of their further elaboration and refinement is a matter of evolutionary detail. For if we assume that the general principles of evolution are true we immediately recognize the cogency of this view. That which we now call the hand has not always been the perfect instrument that it is nor has the ear always been so keenly adjusted as at present. It has required undoubtedly many million of years for the eye to reach its present degree of complexity and adaptability. Yet in all these cases the different organs existed in potentiality from the beginning; the metes and bounds of the hand, the ear and the eye were laid out primordially. Evolution has specialized and adjusted them to environments and needs. Thus it will be seen that while the intuitional faculty was designed for manifestation from the beginning it has nevertheless required ages for its appearance even in the most rudimentary fashion.

Almost the entire content of human knowledge is based upon assumptions or hypotheses; in fact, is but a mass of these, and especially is this true of mathematics, science and philosophy. Of course, there are certain minor observable facts which by reason of the seeming permanence of their existence have been eliminated from the category of assumptions. But even this elimination when it is traced to its depths may be found to be erroneous, and perhaps after all, when we have really begun to know something of the reality of things, may have again to be placed in this category. And then, too, the hypothetical nature of our knowledge is due largely to the Thinker's method of contacting the objective world which is the subject of his knowledge. It is because it is necessary for him to interpret the neurographical symbols which sense-impressions make in the brain matter according to a psychic code that renders his knowledge of things in general hypothetical. His interpretations are based upon an assumed value which experience has taught him to give to each neurogram. But even when his interpretations leave nothing to be desired in respect to their accuracy of apprehension of what the neurographical vibration implies there is that further barrier to his cognition of reality which is due to his remote removal from the object itself and the consequent extreme difficulty, if not present impossibility, of identifying his consciousness with the essence of the objects which he contemplates.

When the Thinker's consciousness is presented with a neurograph of say, a cube, it is not the cube itself which he contemplates or observes; it is the neurograph or psychic symbol which the sense-impressions make in the brain. His consciousness deals not with the object but with the symbols. It is true that when he verifies one neurograph by another, as the *scopographic* or sight impressions by the *tactographic* or touch impressions it is found that the delivery thus determined is a true enough representation. It is also true that the Thinker, as a rule, does not accept a neurograph as valid until it has been verified by at least one or more presentations through his outer sense organs. It occurs, therefore, that all such deliveries are verified and corrected by one or more sense witnesses before final acceptance by the Thinker; but even then it cannot be said that his notions thus gained are in all respects correct and true to the standards set up by the brain-consciousness not to mention higher forms of consciousness. And then, when we consider that in addition to the numerous chances for error which naturally inhere in this method of cognition it must also be apparent that the Thinker's approach to the reality of things is much impeded by his separation therefrom, the unreliability of our ordinary methods of cognition is much emphasized.

But aside from the egoic or brain-consciousness there is the higher consciousness of the Thinker himself. For the brain-consciousness is merely his method of regarding and comprehending the neurographical deliveries, the psychic code by which he systematizes and organizes his cognitions or impressions of the sensible world. This higher consciousness constitutes the faculty *a priori* for the Thinker on a higher plane of existence, and because it deals with elements altogether unlike those which make up the data of brain-consciousness is, accordingly, less liable to error in its judgments of the supersensuous presentations than is the objective or brain-mind. In fact, it is difficult to conceive of a state or conditions wherein, supported as this view contemplates, the intuition should err in judgment. Viewed from the standpoint of external impedimenta, this condition may be said to be due to the absence of sensuous barriers which would otherwise prevent the near approach of the Thinker's consciousness to the essence of things which is the object of his consciousness on this higher plane. Directly, however, it is undoubtedly due to the fact that, following the lead of life itself, yea, as the veritable handmaid of life, it cannot err where life is concerned. When dealing with notions *a priori* or intuitograms the Thinker is relieved of the onerous necessities and limitations incident to the examination and determination of neurographic symbols registered in the brain cortex and so is free to study, to examine and judge at first hand the impressions which are received from his own plane of intuition. The difference is about the same as that which should exist between the methods of communication between two telegraphic operators when in one instance they would have to depend upon the deliveries conveyed over the wires, while in the other, when they stood face to face with each other, they could communicate by direct conversation. In the one case the method of communication is direct and simple, while in the other it is indirect, circuitous and complex. It can, therefore, be readily seen that in all cases where the approach is made in a direct, simple manner the probability of error is much less than in cases where the intellectual approach is less direct and more complicated. Hence in drawing conclusions as to the relative importance of the two mechanisms of consciousness, the *a posterioristic* and the *a priori*, it is necessary to bear in mind the comparative superiority of the one over the other as a means of cognition. It matters little that the intuitional faculty is not so well developed as the tuitional because it is but natural that inasmuch as the Thinker's needs are subserved in the sensuous realm by the tuitional consciousness it should, from more active use, gain somewhat over the intuitional in facility of expression and general utility. And the more so, because the two faculties serve different purposes; one is attuned to receive impressions from a subtler plane while the other is fitted for contact with the phenomenal universe; one is related to the conceptual while the other is related to the perceptual. They differ not only in function, but in nature as well. There is, of course, a natural barrier consisting of the inherent limitations of each faculty which prevents the full mergence or unification of the two states of consciousness so that there exists a state of consciousness the data of which the brain-mind is unaware, it being able only at rare intervals even to receive so much as slight impressions from it in the nature of intuitional flashes or inspirations and the like. Viewed in this light it would appear that the cognitions which are most truthworthy are those which are presented by the intuitional faculty because they are nearer to the essential reality of things; they have to do more specifically with the nature of that which *appears* while the tuitional mind can only regard that which is the appearance. Herein lies the whole difference.

The natural outcome of this division of labor between the tuitional and the intuitional is the establishment of the fact of man's relationship both to the phenomenal and the real; that in his psychic nature must reside the faculty of apprehending the real and that he shall one day awaken into activity this now latent faculty whereby he may make a direct and naked contact with reality.

If it be true that, as PLATO said, God does geometrize, and that the divine geometry, as will appear, is based upon a system, an alphabet which taken together are the point . , the line ——, △, the square □, and the circle ○, then, we have in this geometric alphabet the very secret of the divine geometry. With these, and in the kosmic laboratory of *chaogeny*, the Creative Logos has measured off the limits and confines of space; with them He has traced out its dimensions the archeological evidences of which we may view in the space-mind itself; and with them he has established the manner of its appearance to the Thinker. In dimensions, three, and yet not three, but one, Space, the eternal progenitor of all forms and energies, having received the divine fiat in the beginning that thus far it should extend and no further, persists in faithful obedience to the law of its being— tridimensionality. It must be so because it is thus sanctioned by the highest faculty in man that can render judgment thereon. If tridimensionality inhere in the space-mind, as the law of its being and in the intuitional consciousness as the norm of its essential nature and as the easiest and simplest expression of the tuitional mind, how can it be gainsaid that these considerations obviate the necessity of the mathetical hyperspace?

If the reality of things is hidden from us and if we are, therefore, unable to perceive their real essences it is because our mode of thought and our consciousness have obscured our vision and limited us to this state of paucity of perception. It is not because reality is itself a hidden, inscrutable quantity nor that its *modus vivendi* is "unknowable"; but because we being multiformly limited, "cabined, cribbed and confined" are resultantly lacking in the power to discern that which otherwise would be most obvious to us. It may well be set down as axiomatic that when, in the process of our thinking, we arrive at the inscrutable, the unknowable and the infinite, it is evident that our thought processes are dealing with a form of realism which is higher and beyond the possibilities of our loftiest thought-reaches. And in order to symbolize to itself this condition the intellect poses such terms as "inscrutable," "unknowable" and the "infinite" simply because that is the best it can do. Hence when it is said that space is infinite it is apparent that the mind recognizes that when it contemplates space it is dealing with something whose degree of realism transcends its powers of comprehension. Infinity is a relative term, and in fact, decreases in extensity in the proportion that the consciousness expands and comprehends. It is not unlikely that should the intellect one day discover that it had awakened into union with the space-mind it would immediately reject its preconceived notion of the infinity of space. But we need not wait until the coming of this far off event in the path of psychogenesis; for we can here and now perceive with what must be a higher faculty than the intellect the verity of this conclusion.

But certain it is that the intellect, in the pride and arrogance of its traditional heritage, will not without a great struggle yield the ground and prestige it has held for an aeon of time; and in vain does the intuition serve notice of dispossessal in these premises; but however stubbornly fought the battle, however tenaciously held the position time will discover the weakening of the intellect's hand. Death for the intellect may ensue as a result of the conflict but it will be a death wherefrom it will arise, quickened, revivified and uplifted by its disposer, the intuition, upon the remains of its dead self to a higher and grander state than it has ever enjoyed before.

Space is not static. It is dynamic, potential and kinetic. It is a process, a becoming. Its duration as a process is never ending. Its extensity is limited and finite. The so-called infinity of space is one of the capital illusions of the intellect which can only be removed by an expansion of the consciousness, by a mergence of the individual consciousness with the space-consciousness. In the ever-widening circle of the individual consciousness lesser realities give way to greater as the darkness recedes from the light—the lesser appearing in comparison with the greater, as the consciousness broadens, as matter to spirit, as night to day or as limitation to non-limitation. Thus the most solid facts and conditions of our limited life are but the shadows of the deeper realities which shall be revealed to the Thinker in the days of his larger and more glorious life of freedom from limitations.

And now it will appear that the whole fabric of our knowledge shall have to be reduced to the bare warp and woof; for nothing is real but these. It is as if the Thinker, using the tuitional mind, had been in all times past studying the design woven in the surface of a very thick plush carpet. There are the warp and woof, the long vertical threads which make the plush and the intricate design appearing on the surface. Our knowledge may be likened to the design. It represents the contents of our knowledge. We have not even so much as begun the study of the nature of the vertical threads as they appear beneath the surface to say nothing of beginning the study of the warp and woof. The warp and the woof are the realism of the kosmos; the vertical threads are the roots and stem of the phenomenal world; the design is our sensible world as it appears to the intellect. The life of the intellect has been spent in contemplating this design; while of the hands which wove the carpet, of the mind which directed the hands and of the spirit which vitalized all, it knows nothing nor indeed can it know anything. Where shall we say are those hands, that mind and that spirit which made the carpet possible and an actuality? In vain do we search among the remains, among the soft, glistening threads of the carpet or among the intricacies of the design. For they are not there. They have passed on. The intellect looks at the design or at the vertical threads and because it is unable to follow them to their source, it decides that they are infinite, inscrutable and unknowable. But not so. All that is required are eyes to see and a mind (or shall we say a mind vitalized by the intuition) trained to discern the threads as they point upward with their termini firmly rooted in the warp and woof of the fabric. But we must first master the design, and then turning to the threads, master them. Then shall the doors of kosmic reality swing wide and the Thinker shall be ushered into the eternal palace of kosmic realism wherein he shall find the great secret, the heart, the purpose, the beginning and the end, the very nature of things-in-themselves.

The nature of every degree or condition of realism is so constituted that its qualities, characteristics and limitations are exactly adequate for the satisfaction and fulfillment of all the requirements and needs of every possible state of normal consciousness. So that each degree of reality and each state of normal consciousness is sufficient and complete in itself and mutually satisfies the necessities of each other. The substratum of reality or life which extends from the heart of the kosmos to the extreme limits of the phenomenal universe exists in degrees, not discrete, but continuous. And these merge into one another by insensible stages. Such is the imperceptible continuity of the whole as each degree is gradually immerged into the other that only the limitations of consciousness itself make it to appear as if it were discontinuous. For every stage of realism there is a state of consciousness which answers to it completely and sufficiently. So both the state of consciousness and that of reality, manifesting at any given stage, seem to be complete and final for that stage. Realism or life and consciousness possess only a relative finality fashioned upon the necessities and requirements for any given state of being. Consciousness alone fixes the apparent limits of life; it also determines the state of our knowledge of life. And thus when the Thinker is confined to any stage of reality and congruent degree of consciousness it appears that what he there finds is ample for all his purposes. Accordingly he is convinced that that stage is the final consideration of his scope of motility. It is only when he is able to raise his consciousness to a point where he can contact higher realities that he becomes aware that there are higher stages in which his consciousness may manifest. This peculiarity of the Thinker's consciousness is accentuated when he allows himself to become wholly engrossed with a study of the phenomena of that stage in which he can consciously function. Hence it constantly occurs that men exhausting the study of the phenomenal find themselves floundering upon the beach of the outlying shores of consciousness where in sheer desperation they fall into the illusion that they have indeed reached the limits of manifested life and that beyond those limits there is no organized being. Unconscious are they that in ever widening circles the fertile lands extend and await the awakening of their consciousness when they may till the fallow ground of this new domain and begin again the search for the ultimately real.

With respect to the present powers of consciousness, it cannot be successfully controverted that the concept of tridimensionality of space is sufficient for all purposes. It must be so for it is not only an aspect of the phenomena of space but of reality as well. This fact is attested by the nature of mind that answers to the nature of space. Tridimensionality characterizes the entire extent of consciousness and life, and therefore, of space itself. This characterization may be traced to the very doors of the heart of space where the three become one. Nor would this conception be in the least vitiated if it were allowed that the mass of the phenomena of the supersensuous world, lying in close proximity to the sensuous world, does present itself to the consciousness in a four-dimensional manner and that the phenomena of a still higher plane present themselves in a five or n-dimensional manner to that state of consciousness which may be congruent with them; because then we should be making allowances for the changes in phenomena and their mode of presentation to the consciousness which by no means implies a corresponding change in reality or life. All phenomena are fashioned by the intellect. The phenomenal world is just what the intellect interprets it to be. It is that and nothing more. Its qualities, attributes and characteristics are such as the consciousness gives to it. It exists only for the purposes of the evolving consciousness. And, as an instrument of consciousness, its existence is strictly subject to the evolutionary needs thereof. In that moment that the immediate needs of the consciousness shall no longer be able to find satisfaction in the phenomena of any plane of nature, in that moment the phenomena of that plane disappear, recede and are swallowed up in the maelstrom of eternal reality.

In the gradual expansion of consciousness as it passes through the infinite series of grades of awareness meantime becoming deeper, broader and more comprehensive as it proceeds, there may be observed running through all these planes and orders that which is neither the phenomena of the various planes nor the consciousness; but which must be the substructural basis of both, remaining the same, unchanged and unchangeable. That is the thread of reality, the passage of life itself which is the eternal basis of all. Now it is to this reality, life, that the space-mind is related and in which its roots, its heart and the very center of its being are at one with the divine mind of the kosmos.

The question of dimensionality is solely a concern of the objective or brain-mind which is the intellect. It is one of the ways in which the intellect endeavors to understand phenomena. It is an arbitrary contrivance devised by the intellect for its convenience in studying the world of things. Without it, as obviously appears, the intellect would not be able to go very far in its consideration of the minor problems which inhere in material things. The fourth dimension is but another attitude, another contrivance, which the intellect has devised in order that it may study from another angle the evanescent phenomena of the world of appearances. Having apparently exhausted the possibilities of motion in three dimensions, and being driven on to the acquirement of more picturesque views by the very necessity of its continued growth, it has betaken itself to another higher mountain peak, called "hyperspace" where with larger lenses and higher powered instruments it is beginning to scan the landscapes of a new intellectual realm of consciousness. Yet the celestial wonders of this new-found realm of consciousness remain in undisturbed forgetfulness or neglect. But it is not by a scrutiny of mathetic landscapes nor by a study of the celestial wonders that the Thinker shall one day realize the object of his eagerly pushed quest after the real; for he shall find it, if at all, in the temple of the kosmic mind which is not made by the intellect nor meted and bounded by geometric systems of space-measurement.

In all the learned pother incident to the mastery of the phenomenal, the furniture of the world of the senses, it is as if the self in man, the Thinker, sat secluded in a six-walled tenement, and hence six times removed from the subject of his study, and endeavored to interpret that which appeared to his vision. And thus, thinking that what he sees is the only reality, he remains in inglorious nescience of the reality of that upon which he himself stands, unconscious that the tunnel-shaped aperture through which he peers leads not outward, but *backward* and within to the habitation of the real of which he himself is a part. Men are deeply and well-nigh hopelessly concerned with appearances, with static views of life, with instantaneous exposures. Life, reality and all the eternal verities pass on and assume countless postures, attitudes, moods, tenses and nuances. The intellect is content to occupy itself with a single tense or mood. Indeed, it has no aptitude or power to consider more than one at a single time. It thus misses the continuity, the ceaselessness, of life. What is more, every singularity, every attitude, mood or tense which the intellect grasps for consideration is immediately remade so as to fit its own moods and tenses. And upon each and every nuance the intellect immediately imposes its own form—actually and literally rehabilitates them with its own habiliments. Unfortunately, this peculiarity occludes the intellect from any approach to the true nature of that phase which it can grasp.

Hyperspace is one of the illusions of the phenomenal; it is the dress which the intellect has superimposed upon a single nuance; it is a mask which is an exact replica of the mood of the intellect. Yet through this mask the intellect grandly hopes to approach reality. The period through which the mind is now passing is a repetition of the evil days of scholasticism when men set out to determine the exact number of celestial beings that could be perched upon the extremity of a needle point. It is a time when men's minds easily assume grotesque and hideous shapes and their thoughts become the embodiment of fantastic entities. The exclusive occupation of such minds becomes the fabrication of mathetic monstrosities which rapidly deliquesce upon the first approach of the real or the appearance of the first ray of intuition which may escape through the dim and misty condition of the intellectual over-hangings. It will not be ever thus; for the Thinker will one day pass from a study of the arrangement of phenomena in space and by well-ordered steps come once again to himself. And then through the maze of it all set out upon the true path——the tridimensionality of space following which he will inevitably approach the citadel of the real, the kosmic space-mind.

CHAPTER VII

THE GENESIS AND NATURE OF SPACE

Symbology of Mathematical Knowledge—Manifestation and Non-manifestation Defined—The Pyknon and Pyknosis—The Kosmic Engenderment of Space—On the Consubstantiality of Spatiality, Intellectuality, Materiality, Vitality and Kosmic Geometrism—Chaos-Theos-Kosmos—Chaogeny and Chaomorphogeny—N. MALEBRANCHE On God and the World—The Space-Mind—Space and Mind Are One —The Kosmic Pentoglyph.

Geometry is concerned primarily with a study of the measurement of magnitudes in space. Three coördinates are necessary and sufficient for all of its determinations. Metageometry comprehends the study of the measurement of magnitudes in conceptual space. For its purposes four or n-coördinates are necessary and sufficient. Perceptual space is that form of extension in which the physical universe is recognized to have been created and in which it now exists. Conceptual space is an idealized conception belonging to the domain of mathesis and has no actual, physical existence outside of the mind. Mathematical space represents the idealism of perceptual space.

Geometrical magnitudes may be defined as symbols of physical objects and geometry as a treatise on the symbology of forms in space. In fact, all cognitive processes are simply efforts at interpreting the symbolism of sense-deliveries; and the difference between mere knowledge and wisdom, which is the essence of all knowledge, is the difference between the understanding of a symbol and the comprehension of the essential nature of the thing symbolized. So long as knowledge of space is limited to the understanding of a symbol or symbols by which it is presented to the consciousness so long will it fall short of the comprehension of the essential nature of space. In vain have we sought in times past to understand space by studying relations, positions and the characteristics of forms in space; in vain have we based our conclusions as to its real nature upon the fragmentary evidences which our senses present to our consciousnesses. It is as if one had busied himself with one of the meshes in a great net and confined his entire attention to what he found there, meanwhile remaining in complete ignorance of the nature of the net, how it came to be there, of what it is made and how great its extent may be.

There is ever a marked difference between a symbol and the thing which it symbolizes. Words are the symbols of ideas; ideas, as they exist in the mind, are the symbols of eternal verities as they exist in the consciousness of the Logos of the universe. There may be a wide diversity of symbolic forms which represent one single idea; as, for instance, the variety of word forms which represent the idea of deity in the various languages. Likewise there may be a multiplicity of ideas which represent a single verity. But neither is the idea nor the word the real thing in itself. That quality of a life-aspect which we call its *thingness* has an essential nature which cannot become the object of consciousness except by virtue of its representation through ideas and their symbolisms, and even then, the thing which we conceive is not the nature of a quality of the life-aspect but an idea of it—a symbol which stands for that idea. In order, therefore, for the mind to arrive at an understanding of an eternal verity, such as space, it must first be able to synthesize all of the representative ideas and then abstract from their compositeness a notion of its essential nature. But this can be done only by identifying the consciousness with the essential being of the object considered. In other words, the consciousness and the intrinsic being of forms, principles, forces and processes must embrace each other in the intimacies of direct cognition; the life which is consciousness and that life which is essential being, being coeval, coördinate and mutually responsive, must in so close a contact as here intimated reach an understanding of the realism shared by both. That is, the human consciousness, following in the wake of life and consisting of a specialized aspect of life itself, will, by such an intimate approach to the life-principle of forms, readily understand; for it has only to recognize a replica of itself in rendering its judgment. But it is not claimed that such a state of recognition by the consciousness of life itself can be attained at all by ordinary means, neither is it believed that it is the next stage in conscious evolution. However, it is not doubted but that such an exaltation of the consciousness is possible, yea practical; but the difficulties which beset the path of attainment in this direction are so great that it may as well be considered unattainable. The mere fact of these difficulties, however, only re-emphasizes the insufficiency of the intellectual method. The identification of consciousness with essential being is a procedure which cannot be accomplished by an act of will directly and immediately. Because it is a process, a series of unfoldments, an adjustment of the focus of consciousness to the kosmic essentialities which constitute the substructure of the manifested universe. In the very nature of things, a kosmic essentiality cannot be viewed as being in manifestation especially in the same degree as ordinary physical objects are manifest. The former is a state, a potentiality, a dynamic force, an existence which should be thought of as an extra-kosmic affair dwelling on the plane of unity or kosmic origins; while the latter are the exact opposite of this. The one can be seen, felt and sensed while the other is the roots which are not seen but lie buried deeply in the heart of the universal plasm of being and beyond the ken of sensuous apperception. The term *manifestation* is both relative and flexible in its use. It is relative because it will apply equally to all stages of cognition. A thing is in manifestation when it is presentable to the ordinary means of cognition belonging to any stage of conscious functioning; it is not in manifestation when it is beyond the scope of the Thinker's schematism of cognitive powers. Its flexibility is seen in its ready yieldance to the entire range of implications inhering in the process of cognition, fitting the simplest as well as the highest and most complex.

Great is the gulf which is interposed between manifestation and non-manifestation; and yet the two, in essence, are one. They are linked together as the stem of a flower is joined to its roots. Likewise one is visible, palpable while the other is invisible, impalpable though no less real and abiding. As the thin crust of earth separates the stem, leaves and flowers from the roots so the limitations of man's consciousness separate the manifest from the unmanifest. Similarly, as when the surrounding earth is removed from the roots and they are laid bare revealing their continuity and unity with the outputting stem and flowers, so, when the limitations of consciousness are removed by the subtle process of expansion to which the consciousness is amenable so that it can encompass the erstwhile unmanifest, it, too, will reveal the eternal unity of the kosmic polars—manifestation and unmanifestation.

There is but one barrier to ultimate knowledge and that is the human consciousness however paradoxical this may seem. The unutterable darkness which shuts out the so-called "unknowable" from our cognition is the limitation of man's upreaching consciousness. These limitations constitute the difference between the human intellect and the mind of the Logos. Nevertheless the outlying frontiers of man's consciousness gradually are being pushed farther and farther without. Every new idea gained, each new emotion indulged, each new conception conquered, and every mental foray which the Thinker makes into the realm of the conceptual, every exploration into the abysmal labyrinth of man's inner nature are the self's expeditionary forces which are gradually annihilating the frontier barriers of consciousness and thus approaching more closely upon the *Ultima Thule* of man's spiritual possibilities.

Space is in manifestation. It exists and has being whether it is viewed as an object, an entity or the mere possibility of motion. That it offers an opportunity of motion and renders it possible for objects to move freely from point to point cannot be denied and yet this fact has no bearing whatever upon the essential nature of space. The very fact of its appearance, its manifestation, makes it obvious that it is the nether pole of that eternal pair of opposites—manifestation and non-manifestation, being and non-being, which are essentially one.

It will be seen from figure 17 that the period of involution embraces seven separate stages, the *monopyknon, duopyknon, tripyknon,* etc., being the unit principle or engendering elements of the respective stages. Involution comprises all creative activity from the first faint stirrings of the void and formless chaos until the universe has actually become manifest and dense physical matter has appeared. It is divided into two cardinal periods, namely, the chaogenic period during which primordial chaos is given its character and directive tendencies for the world age. It is a phase of duration wherein the fiat of kosmic order is promulgated, and consist of three stages, monopyknosis, duopyknosis, tripyknosis[24] gradually, insensibly, gradating into the fourth or *quartopyknotic*; the chaomorphogenic period is likewise divided into three stages—

quintopkynosis, sextopyknosis and *septopyknosis,* developing out of the fourth gradually. The quartopyknotic stage is the stage of metamorphosis or transmutation wherein the transition from non-manifestation to manifestation is completed; it is also the stage of kosmic causation, because from it spring the matured causes or "vital impetus" which engender all that follow.

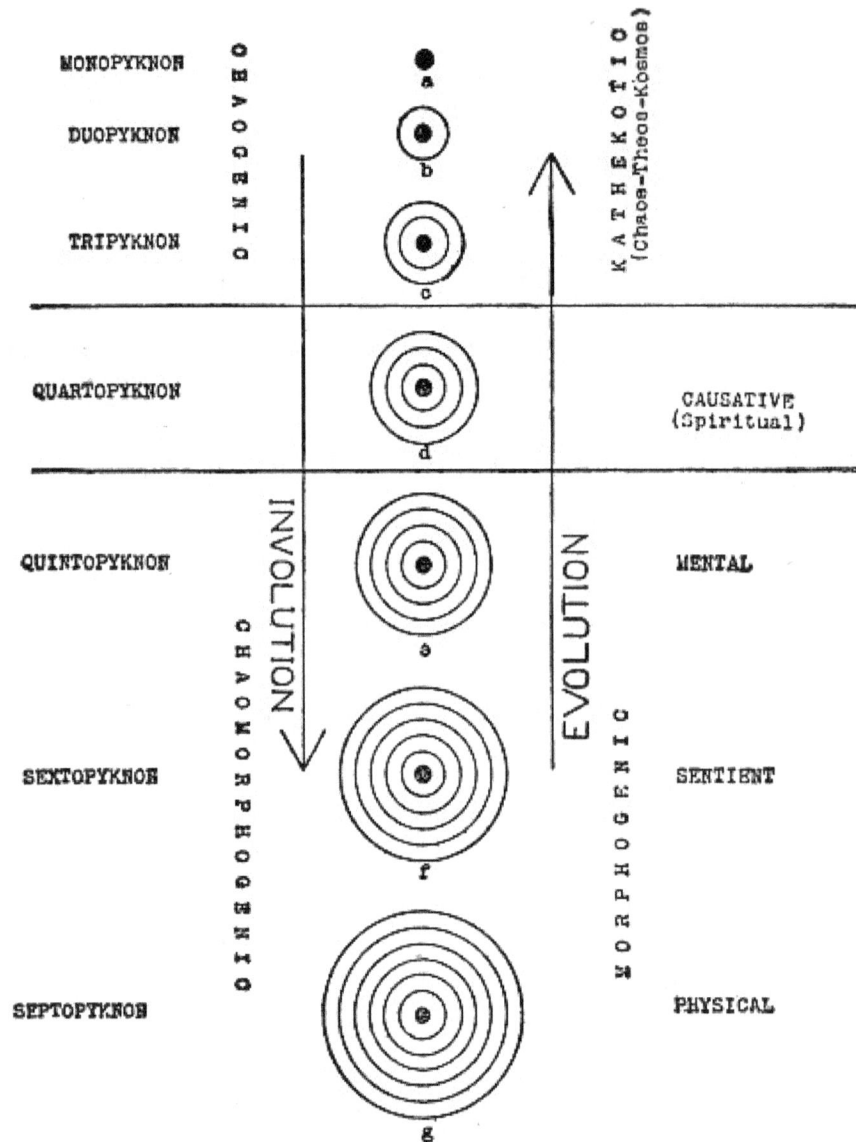

MONOPYKNON

DUOPYKNON

TRIPYKNON

THEOGENIC

KATHEKOTIC
(Chaos-Theos-Kosmos)

a

b

c

QUARTOPYKNON

CAUSATIVE
(Spiritual)

d

INVOLUTION

EVOLUTION

QUINTOPYKNON

CHAOMORPHOGENIC

MENTAL

e

SEXTOPYKNON

MORPHOGENIC

SENTIENT

f

SEPTOPYKNON

PHYSICAL

g

FIG. 17.—Involution and Evolution

The close of the involutionary phase of the world age is marked by the final deposition of dense physical matter and this is closely followed by the beginnings of the evolutionary movement which, like the involutionary movement, is divided into two cardinal periods, namely, the morphogenic (during which are produced, in turn, insensible forms, sensible forms and spiritual forms) and the kathekotic period which marks the perfection, the consummation of the evolutionary movement. These two cardinal periods of the evolutionary phase of duration and the two cardinal periods of the involutionary phase complete the kosmic age, the "Great Day of Brahma." The concentric circles, beginning with the dot and ending with the seven concentric circles, and designated as a, b, c, d, e, f, g, are representations of the constitution of the respective units corresponding to each of the seven subdivisions. They symbolize the seven degrees of condensation or pyknosis which comprise the genesis of space, on the one hand, and on the other, the stages of unfoldment. Because, during involution all potencies, powers and characters were being infolded, involved; but during evolution, these are being unfolded, expressed, evolved.

The figure 18 is another view of these two major movements, involution and evolution. The genesis of space is here shown symbolized by the Kosmic Egg. The seven stages of involution are referred to as, the monopyknotic, duopyknotic, tripyknotic, quartopyknotic, quintopyknotic, sextopyknotic and the septopyknotic; while the corresponding stages of evolution are referred to as, the physical, the sentient, mental, causative or spiritual, the triadic, duadic and monadic, indicating that the principle of physicality is succeeded by the principles of sentience, mentality, spirituality, and the three forms of kathekotic being. This symbolism, it should be stated, is designed with respect to the universe and man and has no reference to other possible evolutions than the human and contemporaneous animal, plant and mineral evolutions.

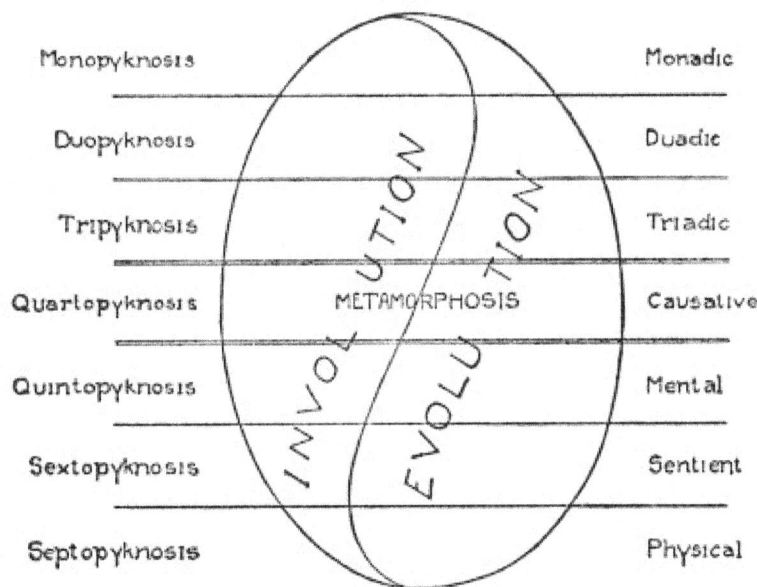

FIG. 18.—The Genesis of Space

To follow the ramifications of the symbolism above would involve a survey of all branches of knowledge, and indeed, would be out of place in this book. Only the widest general outlines can be suggested here and it is believed that this is sufficient to enable the reader to grasp the magnitude of the symbol and to understand its purpose and intent.

It would appear, therefore, that if it is possible for the intellect to traverse by means of a study of kosmic symbolism, used as a standard of reference, the entire length of the bridge which engages the antipodes into an eternal unity, something may be gained in the way of a more definite and clearer understanding of the essential nature of space in its relation to kosmogenesis.

In the diagram, figure 17, is shown a table which represents the stages of space-genesis. It will be noted that the whole scheme is divided into seven stages. It is not an arbitrary division simply but a symbolic one and represents fullness, completeness, entirety. The names given, namely, "monopyknotic, duopyknotic," etc., represent the symbolic characteristics of each stage in its relation to the universe in the process of becoming. The terms "monadic," "duadic," "triadic," etc., are representative of the seven planes of matter in the universe.

A *pyknon* is a kosmic principle and represents the typal aspect of kosmogenesis. It is a generic term and may be identified in its relation to the various stages by the prefix. The monopyknon belongs to the ulterior pole of the antipodes on the side of non-manifestation. So do the *duopyknon* and the *tripyknon*. Pyknosis is a process of kosmic condensation, or limitation for purposes of manifestation. It is a stage in the descent of the kosmic Spirit-Life, a degradation of non-manifestation into manifestation, and is, therefore, the cardinal causative principle of creation. The term pyknon being generic is applicable alike to a particle of matter, a state of being, a condition of existence, a process or a principle. The *monopyknon* is, accordingly, the primary aspect of the process of kosmic pyknosis. It is the archetype and therefore all inclusive and omnipotential. But whether regarded as a singularity or as a whole it should never be divorced in thought from the primal act of creation. It represents the first act of material differentiation in the being of the creative Logos on the plane of non-manifestation. It is the beginning of every great planetary or kosmic *manvantara* or period of manifestation. During either a planetary, solar or kosmic *pralaya*, or gestatory period, the kosmic plasm is in a quiescent, undifferentiated condition. This undifferentiated plasm when acted upon by the will of the Creative Logos, *Fohat*, as He is sometimes called in Eastern Philosophies, begins to become conditioned, begins to differentiate. The primal act or stage in such a process is the formation or appearance of a monopyknon. It then becomes the characteristic aspect of that stage.

Monopyknons are the quiescent, unawakened, though potential and archetypal principles peculiar to the monopyknotic period of space-genesis which are ultimately to become, on the physical plane, singularities of life of whatsoever kind. *Thus the lineage of every single life-form or principle in the universe runs unbroken back from the present of the Now to the present of monopyknosis. So great is the design of the kosmos that the entity which is now man or the atom was started on its journey to this culmination at the break of the Great Kosmic Day when the omni-pregnant wheels of monopyknosis first began to turn.* Duopyknons and tripyknons constitute the two remaining stages on the plane of non-manifestation. And their correspondences in every stage of involution or the descent of spirit into matter are eternal and kosmic. Likewise the lineage of the dual and triple aspects of all life forms on the path of evolution may be traced rearward to the duopyknotic and tripyknotic stages of kosmogenesis.

The metamorphosis by which the monopyknon becomes a duopyknon contrives the differentiation of the pristine plasm of kosmic being so that the first becomes the ensouling or vitalizing principle of the second; and, in turn, the second becomes the vitalizing or inner principle of the third; the third of the fourth and the fourth of the fifth and so on throughout the series until the last is reached which is the *septopyknon*. The septopyknon is, therefore, a seven-principled form. It is both unitary and septenary—unitary in the sense that the seven are really one and septenary for the reason that each of the seven principles, in the course of evolution, becomes a separate process specially adapted to functioning upon its peculiar plane of matter. Thus it is seen that the utmost significance attaches to this septenary pyknosis of the kosmic plasm of life. The implications of this conception are, of course, too vast and multifarious to be set down here. We shall have to dismiss it with one observation only, and that is: *every single appearance of life and form in the totality of such appearances is rooted in kosmic pyknosis where it has received its inner vitalizing force, its form and the law of its mode and manner of appearance together with the metes and bounds of its existence.*

These processes, monopyknosis and duopyknosis, are to be regarded as taking place, each in its own period, everywhere throughout the Body of Being of the Logos but on the plane of non-manifestation. They are states of preparation for manifestation analogous to the germinative period of the seed or the egg. They represent the first stirrings of the kosmic plasm and contain the promise and potency of all that is to succeed them. There is one other stage, coördinate with these two, and that is the tripyknotic which completes the unmanifest trinity and constitutes the archetypal vehicle whence proceeds the manifested universe. The ensouling principle of the tripyknon is the duopyknotic principle. But when the descent has reached the tripyknotic stage the three have been merged into one and the characteristics which were peculiar to each as a separate pyknon are then fused into a single quality having three aspects which are mutually interdependent and coördinate.

The unmanifest trinity, now complete as a result of the triple pyknotic process, is the imperishable and ever sustaining radix, the all-mother of the manifested universe. It is the Golden Egg laid by the god *Seb* at the beginning of a great life-cycle. It is also *Chaos-Theos-Kosmos*, i.e., kosmic disorder, divine will or generating element, and kosmic order or space. In it, as in an egg, resides in kosmic potency all that the universe is to become in any "Great Day of Brahma" or any Great Kosmic Life Cycle. Its eldest born is SPACE, physiological and perceptual, and the latter is the eternal father-mother of the universe. *Space is, therefore, the male-female principle of manifestation; in its kosmic womb all forms are created, developed, evolved and sustained.* Into it again, at the close of the Great Day, all existences, forms and all principles matured and ripened by the vicissitudes of kosmic evolution will be inhaled with the return of the Great Breath of Life. The unmanifest trinity is the archetype, and therefore, the pattern or model for the manifest, embodying in potency all that the manifest may ever become. Forth from the unmanifest proceeds space as a dynamic process endued with the potentiality for generating all that the manifested universe contains.

The terms "unmanifest," "unconditioned" and "unlimited" have a special meaning here and are used in the same sense as the mathematical term "transfinite," and therefore, imply a transcending of any finite or assignable degree or quality of manifestation. Hence they should be distinguished from the term "absolute" which has a different implication. So that although the triple process outlined above may not be viewed except as a characterization of the plane of non-manifestation, and hence of the primordial activity of the Creative Logos, there is nothing in the symbolism to warrant the identification of this process in any way with the Logos in absolution. For on this view Absolute Being is, in a large measure, sacrificed when the monopyknotic process is begun and the monopyknon (kosmic principle) begins to appear. Absolute Being, while it may not be defined, delineated or described may be symbolized by the ideograph: "action in inaction"; "being in non-being"; "manifestation in non-manifestation"; for these are symbols merely and do not describe or delineate.

We have observed the subtle connection which exists between manifestation and non-manifestation and have seen how that, as the roots of a plant sustain the outer growth of stem and flowers, the former being the matrix out of which proceeds the latter, and that in like manner does the unmanifest sustain the manifest; it should, therefore, be clear that the Body of Being of the Unmanifest Logos, in a similar manner, is the basis and cause of all that is manifest or in existence; *and yet it is more, it is essentially all that is and all that the all is to be in manifestation at any time throughout the immeasurable process of kosmogenesis. It is the Self of the Universe, the meta-self of the great diversity of selves.* The Self, however, although it cannot be said to exist except as a simple, homogeneous quantity is nevertheless, in the very nature of things, a triunity, and in essence, not only the basal element of the All-Space and the concrete forms which exist therein, but is also identical in essence and substance with space.

A stage has now been reached in the description of our symbolism when it may be assumed, upon the basis of the foregoing, that the meta-self has become manifest, i.e., in potential and dynamic appearance, as a result of the triple process of pyknosis hereinbefore outlined. The meta-self may then be identified with the Supreme Manifest Deity; for there is ever a subtle identification of the manifest with the unmanifest. Out of their action and interaction the universe is made manifest and phenomenal and is thereby sustained. The reciprocity of action between these two kosmic polars, is the metamorphotic key to creation; it is the symbol of the procedure of Creative Will in the act of creating. It is the transitional process whereby the passage is made from non-being to being; from the unconditioned to the conditioned; from undifferentiation to differentiation and is reflected and symbolized in every natural process wherein matter is transformed from one state to another, or life and mind and spirit diffused, centralized and organized into ever new and higher forms and expressions.

Another important notion to be gained in this connection is the fact that it appears as a logical sequence to the foregoing that the being of the manifested Logos must necessarily fill all space, yea more, is that space in every conceivable essentiality. His limitations are the limitations of space. His qualities, properties and attributes are the qualities, properties and attributes of space and are only different from the original spatial character when manifested through a diversity of forms by whose very inner constitution and exterior form the modifications are accomplished. *All matter in the universe, all energy, and indeed, all manifestations or emanations of whatsoever character, hue, tone or quality are, in reality, His being and nothing but His being. There is, therefore, no form nor ensouling principle whether of life, mind or sheer dynamism which can exist outside of His being and be, even in the slightest degree, absolved from an eternal identity therewith.* Once this idea is grasped and its varied implications noted it then is no longer conceivable that any other order or schematism can be possible in our universe, and that, too, despite the multiform conceptions peculiar to the varied systems of philosophy.

The matutinal dawn of creation came at the close of the tripyknotic movement which resulted in the elaboration of materials, the initiation of principles, processes and types, and the final preparation of the field of evolution. The three processes or aspects of non-manifestation projected in preparation for manifestation, namely, monopyknosis, duopyknosis and tripyknosis represent the earliest stages of germinal development. When these had closed, the Great Kosmic Egg began to germinate; the first faint, indescribable signs of manifested life began to appear. Involution set in. The fourth or quartopyknotic stage, though only slightly differentiated, or rather representing that period of kosmic involution when that which is to become the manifested universe first begins to fall under the sway of kosmic order, is nevertheless the basis of all great world processes. It is just midway between the poles—manifestation and non-manifestation. During this stage the life elements are receiving the imprints of character, being endowed with directive tendencies and stored with such dynamism as will persist throughout the Great Life Cycle in which they are to manifest. It is here that begins the movement of involution, the storing away of those elements and factors, no more and no less, that are to show forth on the upward path of evolution; it is here that matter begins to assume form; electrons, ions and atoms created, or, that those minute processes which on the evolutionary side are to culminate in these are originated. This is the metamorphotic stage. It is the laboratory of the universe wherein *Fohat*, the Creative Logos, prepares the materials out of which and in which the vast diversity of *morphons* or forms is created. Quartopyknosis, accordingly, is the first active step, on the plane of manifestation, which results in the appearance of perceptual space and consequently of physical matter itself as well as all the other grades of matter in the kosmos. Space, brought into existence by the act of the Creative Logos in imposing limitation upon His being, is in its primordial form composed of quartopyknons or quadruplicate principles and tendencies which act in unison and to the accomplishment of a single end or purpose. On this plane or during the continuance of this period of space-creation, the roots of universal law and order are produced. In it are planted the principles of good and evil and a sharp line of demarkation established between all the conceivable pairs of opposites which exist. It accounts for the duality of life and form. Male-female; father-mother; positive-negative; Rajah-Tamas (action-inaction)—all these find in this process their eternal origination. It is the stage of harmony, bliss, ideality, perfection, perfect equilibrium and balance. Here, innumerable ages before they actually appear, the glow-worm and the daisy, the amoeba and the dynosaur, man and the planetary gods alike abode their time awaiting the toppling of the scales of kosmic potency when all would be plunged headlong into the endless labyrinth of becoming.

The quartopyknotic process is similar in all details to the three preceding processes, these latter being prototypes of all succeeding stages of involution. The quintopyknon, accordingly, symbolizes the quintuplicative action of life in its descending movement toward the creation of matter in its densest form. That is, it is a five-fold principle acting in unison and kosmic consistency, infolding in the universal wherewithal that which is to become mental matter on the side of evolution. Just, as may be seen in the diagram, Figure 17, the quartopyknotic process symbolizes, on the involutionary side of the life current that which is to become on the evolutionary side, spiritual essence, so the quintopyknotic deposits the seeds of that which is to become mental matter. During both the quartopyknotic and the quintopyknotic processes all the potencies and promises, residing on the plane of non-manifestation and destined to show forth as spirit and mind are brought into a fuller and more marked degree of manifestation and become the seeds of spirituality and mentality which are to ripen and be ladened with the fruitage thereof many ages hence.

The reader should bear in mind that the processes here described are thought of as taking place at the beginning and as having their roots planted in eternal duration; that they refer to a period long before the universe even resembled anything like its present aspect; at a time even before there were individual minds to perceive it, before even the gods—solar, planetary, super-solar and super-planetary were in existence to take part in the matutinal ceremonials of creation's vast hour of stillness. The mind must accustom itself to go back of appearances, back of time, back of space itself and discern the foundations of time, space and appearances being laid and to perceive that which is no less than the action of the Supreme Deity Himself in brooding over the primordial formlessness which is Himself, and from which will gradually evolve the universe of qualities, conditions and appearances.

The quintopyknon is, therefore, the base of the mind-principle in the kosmos. All the qualities of mind whether in man or in the planetary gods, whether in the moneron or the tyrannosaur, in the mountain or the oak, reside in kosmic potency in the quintopyknon.

NICHOLAS MALEBRANCHE,[25] in one of his very lucid moments, beheld the essential character of the symbology of space and was led to the conclusion that God is space itself. To him it was equally certain that all our ideas of space, geometrical or purely physiological, as well as our notions of the great suprasensual domain of ideas, exist in the kosmic deiform, or body of the Logos of Being. He saw "all things in God." God did not create ideas; they are a part of God Himself; God did not create mind; it is a part of Himself; no kind of matter did He create; it is a veritable part of Himself and indissoluble from Himself. The great outstanding implication of this philosophy is that our consciousness of God is but a part of God's consciousness of Himself; our consciousness of self and the not-self is but God's consciousness of these things. There is no existence of anything, either of the self or the not-self, except in this consciousness. It is refreshing, therefore, to note that although the approach is made from another and entirely different direction, almost the same conclusion as to the ultimate resolution of all chaogenetic elements into what is the very systasis or consistence of the great kosmic deiform, is reached.

But a marvelous vision comes with the dawn of this truth upon the lower mind. It establishes clearly the truth of KANT's notion when he said: "Since everything we conceive is conceived as being in space, there is nothing which comes before our mind from which the idea of space can be derived; it is equally present in the most rudimentary perception and the most complete." The mind cannot get away from the conception of space, because, out of the very essence of space, as a result of the quintopyknotic process, it was produced, created and organized. The idea of space is, therefore, not derived from things in space nor from their relations in space. *It sprang up with self-consciousness.* As soon as the Thinker became conscious of himself he became aware of space. The very state of self-consciousness implies space. The self in man is a specialized aspect of space. Indeed, it is a projection of space. The moment the self can say: "I am," it also can complete the declaration by saying: "*I am Space.*" When the self looks out from his six-walled cabin of imprisonment into the immensity of what we call space he looks out into that which is himself and his immensity; he perceives the source and the ever-present sustenance of his being and recognizes his identity therewith, provided he does not allow himself to become entangled in the philosophical difficulties which the intellect is prone to throw around the simple, yet marvelously complex, notion of self-consciousness.

This should settle, once for all, the question of *apriority*. The *a priori* inheres in quintopyknosis or kosmic psychogenesis. It is the essential nature of mind; it is the mind's lines of organization; it is the law of the mind's being and action. All mental perception originates from things in space. No thought of any detail, of any state or condition, whether limited or unlimited, related or isolated can be conceived except it be of things in space. And this is so, because *mind and space are one*. It is not so with our conception of time. Time is merely an aspect of consciousness in its limitation and does not inhere in the mind in the same manner that does space. In fact, it is not a part of the mind's nature as it has been shown that space is. It would, therefore, seem to be a grave mistake so to coördinate the two notions. Space is the progenitor of mind and is continually identified with mind. Time is the child of consciousness. That is, it is one of the illusions of consciousness which the ego will shed as his consciousness expands. Duration alone is a coördinate of space.

The mind now recognizes space as something apart and separate from itself only because of its unconsciousness of the identity existing between it and space. Just so, it is not by mind alone that the *at-one* state of consciousness shall be attained; for although in one form or another it is able to gain some knowledge of the apparent oneness of all life it cannot directly realize this oneness. In order to do this fully it must be able consciously to identify itself with the life, feel what it feels and experience what it experiences and otherwise come into a conscious relationship with the root and source of life. Space-consciousness is a simple, direct cognitive process; while time-consciousness is a complex, and therefore, indirect process. The former cannot be analyzed. That is, no analysis is necessary to its sufficient comprehension; the latter must always be analyzed and categorized for its sufficient apprehension. Every moment of time whether past, present or future, when presented to the consciousness, is determined by its relationship to some other moment of time. Space is indivisible; time is divisible. Space is an intuitional concept; time is an intellectual concept. Time belongs to phenomena. Space is the root and source of phenomena. Time is the leaves of a tree while space is the life of the tree.

Space-consciousness, in its relation to the present status of mind-development, is itself an illusion; for despite the fact that the Thinker's apperception of it as a state is simple, direct and fundamental, it is only so because of the inability to realize to itself the unity of the seeming two. The attainment of space-consciousness or the space mind, which contrives the understanding of the identity of mind and space also annihilates the consciousness of space as a separate notion from the mind. Once the Thinker's consciousness has arisen to that state where it perceives its unity with space all sense of separateness is lost. Just as when two molecules of hydrogen uniting with one molecule of oxygen to form a new compound lose their identity in the new realization of unity, so does the consciousness when by the alchemy of psychogenesis it becomes identified with space, not only lose its identity as such, but also any consciousness whatever that space exists as something separate and distinct from itself. Imagine the whole of the duration aspect of kosmogenesis crowded into an infinitesimal instant and the bulk of all matter, suns, stars, worlds and planets, condensed into a space less than the magnitude of an hydrogen ion and in this way a symbol of what it may mean to attain unto absolute knowledge or unto the space-mind, may be obtained.

Recurring to the process of quintopyknosis, it may be noted that the quintopyknon or five-fold kosmic principle of life which we have seen to be identical to the seeds of mental matter brought into existence by the reaction of Fohatic energy or the Will of the Creative Logos upon the substance of the quartopyknotic stage, is, symbolically speaking, more dense and compact than the pyknons of the preceding stages. It is ensouled by the quartopyknon. It is a rather complex state of ensoulment consisting of four condensations or pyknoses.

The next stage in the process of kosmic involution which is also concerned with the preparation of the evolutionary field is that of sextopyknosis and implies the senary condensation of the original world-stuff with the view to the formation of emotional or "astral" matter. Identically the same process of ensoulment or involution obtains upon the plane of sextopyknosis as have been observed to obtain upon the preceding planes of involution. Involution must necessarily precede evolution. That which has not been involved, enfolded or ensouled cannot be evolved or unfolded. Whatever potencies, powers and capabilities or qualities and characteristics that may appear at any time in the universe of life and form must have first been involved or enfolded or else they could not have been evolved. Space itself is an evolution. It is a process of becoming, of unfolding, of flowering forth. As it evolves more and more there will appear new and added characteristics and qualities of life and form. New possibilities will arise and in the end a supernal vision of a glorified universe will burst into view.

The scheme of space-genesis is completed during the septopyknotic process wherein the basal elements of dense physical matter and its various gradations are produced and given character, form and direction. But this completion means merely a temporary estopment in the process of kosmogenesis which actually results in the formation of physical matter in its crassest state. It does not mean a final arrest of the entire process which is conceived of as continuing only in a regressive manner back to a kathekotic[26] condition wherein it embodies the fruitage of the entire scheme. The septopyknon, accordingly, is a seven-ply pyknon in which are embodied, in varying degrees of manifestation and phanerobiogenic (life-exhibiting) quality, the essentialities of all that has preceded on all planes and during all stages of space-genesis. That is to say—in the physical life of the universe is confined the essence of all the series of grades of life in the kosmos. In man's physical body are wrapped up all the glories attainable in his long, almost unending pilgrimage of evolution; in it are stored all the possibilities of the spirit; all powers, all qualities, all characteristics, ever intended for man's attainment are in the physical. But they must be evolved, they must be unfolded and expressed. The physical must be *glorified, spiritualized, deified*. For by the way of the glorification and spiritualization of the flesh man may attain unto oneness with the divinity in himself and consequently with the divine life of the world.

To summarize: The genesis of space embraces seven stages, namely, the monopyknotic, the duopyknotic and the tripyknotic which belong to the plane of non-manifestation and are the primordial world-stuff and together make up the unmanifest body of the Logos of Being. These become the seed-germ of the universe of spatiality. The quartopyknotic is the fourth stage in the process of space-genesis and is the *metamorphotic* or crucial stage during which non-manifestation is metamorphosed into manifestation. In it the unmanifest becomes the manifest. It corresponds to the plane of pure spirit, and indeed, embodies within itself all the qualities which spirituality is to show forth during the life of the kosmos. The quintopyknotic is the fifth stage and corresponds to the mental plane, embodying in itself all qualities of mentality in the universe and furnishing the basis and essence of that which is to become the kosmic mind in manifestation. The sextopyknotic is the sixth process and symbolizes the sixth stage which embodies all the characteristics and properties of emotional matter in the universe and is the basal element of the plastic essence of sentient existence in the kosmos. The septopyknotic is the seventh and final stage corresponding to the physical plane of the kosmos and contains in its seven-fold constitution the seeds and potencies of all the preceding stages, as well as all the characteristics and properties which physical matter is destined to show forth during the *manvantara* or world age. These seven processes result in the dynamic appearance of space, the mother and container of all things, and complete the involutionary aspects of kosmogenesis. Evolution began where involution ceased and will end for this *manvantara* when the last vestige of those powers, capabilities and potencies which were involved shall have been evolved unto kosmic perfection.

The measure of the Great Kosmic Space-form was sealed at the close of the involutionary movement of the Great Life Wave. Then its metes and bounds were fixed by the fringe of kathekosity which circumscribed it.

If it be true that the reader found it extremely difficult to grant the connotations of the symbolism when the mental or quintopyknotic stage was reached when illative cognizance was given to the fact that space is also composed of mental matter, it may be still more difficult to grant the claim that physical matter is also essentially a part of space. But this is the implication. *And, therefore, it follows that all matter, all energy, all life and all mind wherever it may be found in the Great Space-Form is space itself and nothing but space.* Hence, it appears that space is indeed the dynamism of the universe. In its kosmic womb the great world egg was formed of its own substance solely and in it still the universe of form persists and evolves withal.

If it be suitable for the physicist to talk of gravitation, electricity, magnetism and force let him do so, for these terms serve the present category of human knowledge; but the human mind will not lament the day when it comes to recognize that these things, these forces, these aspects are nothing more than space-activities and space-phenomena. If a planet's place be preserved in space it is because space, vital, dynamic, creative space, sustains it and from its gentle, yet eternally firm grasp there is no escape. All that the planets, suns and worlds are and all that they may ever become in this *manvantara* or world age have been derived from space, yea, are of the very essence of spatiality. If the chemist choose to talk of chemism, negative and positive, of combining properties and dissociative phenomena let him also become aware that these phenomena are but the external aspects of the inner and ephemeral life-processes of space-forms and that ultimately these, too, may be traced back into an eternal originality within the bosom of the all-mother, spatiality.

Dense physical matter, such as constitutes the physicality of celestial bodies in its ultimate dissociation would, accordingly, be resolved into the original chaogenetic formlessness which marked the chaogeny of non-manifestation although it would naturally be orderly and progressive passing through the seven stages, septopyknosis, sextopyknosis, quintopyknosis, etc., until the end had been reached, meanwhile exhibiting in each plane the phenomena peculiar to the dissociative processes thereof. On this view space is a *plenum* of matter of varying degrees of intensity, ranging from the densest physical to the most tenuous and formless matter of the highest levels of the manifested universe. But as neither the dense material forms nor the other grades of matter have an eternally enduring quality, being alike subject to mutation, space likewise falls under the law of becoming whereby it, too, must yield to the edict of kosmic disorder.

Some may be inclined to argue that since space and mind are one and the same thing it must necessarily follow that whatever possibilities of measurement may be found to exist in the mind would logically be found to exist in space; and that since all the necessary conditions of hyperspatial operations are proved to be existent in the mind the case of the hyperspatiality of perceptual space is proved thereby. In other words, if the fourth dimension can be proved to be mentally construable it is also possible in perceptual space. But these hypotheses are not granted, and neither will they be acceptable to those minds who choose to take that view when it is known that there is a marked difference between the mind that is purely intellectual and mind that is purely intuitional or mind *a priori*. The intellect is fashioned for matter only; it is so constructed as to fit squarely into every nook and cranny, every groove and interstice in matter; yet for the generating element, life, it has no aptitude nor suitable congruence.

The attainment of the space-mind or kosmic consciousness would then imply a mastery of all fundamental possibilities pertaining to all degrees of matter. Thus by becoming conscious in the matter of all the planes one makes a certain definite approach to this ultimate state of consciousness until all the barriers between ordinary self-consciousness and the consciousness of the space-mind have been entirely obliterated.

Pyknosis, in all of its septenary aspects, is concerned primarily with involution or the preparation of the chaogenetic elements for the work of kosmic evolution. It may be thought of as being divided into two great divisions, namely: *chaogeny* and *chaomorphogeny*. It is concerned with the organization of chaos, the establishment of kosmic geometrism in the formless, void, arupic substance and preparation for evolution. Chaogeny, of course, is that kosmic process by virtue of which space itself becomes manifest and in which there is no established order. Chaomorphogeny (from *Chaos + Morphe + Geny*) signifies the activities of the creative Logos in laying the foundations in primordial space-matter of the various star-forms, including nebulæ, worlds, planets, suns, etc., of which Canopus, Jupiter, Fomalhaut and Sirius and our own sun are examples, giving direction and general tendence to their varied life-processes. Both these processes are concerned with the preparation of the field and its consequent fertilization in anticipation of its cultivation and harvest. These two constitute kosmic involution or the great life wave's passage on the downward arc of the Great World Egg or Circle. It is during the chaomorphogenic cycle that the constitution of the universe of manifestation is promulgated; when laws for its government during that *manvantara* are sketched out in the world of nascent spatiality; when the archetype of every imaginable or possible form is projected upon the impregnated screen of creation, then folded in, pushed toward the center, involved, awaiting that time when the life wave begins its passage upon the upward arc and evolution ensues, calling forth all that has been enfolded in the bosom of the pyknotic centers of manifestation. It is easily conceivable that here during the troublous times of the chaomorphogenic enfoldment the now known six directions of space were among the eternal edicts of space-genesis and that that law which now makes it appear that three coördinates, and only three, are sufficient for the determination of a point position in space was imprinted in the very nature of that which was to become space.

The kosmic field having been prepared as a result of the chaomorphogenic activities, lowly and scarcely organized forms begin to appear and the ascent upon the upward arc of the Great Cycle commences. Evolution begins. Its scope is likewise divided into two great stages, namely: (*a*) Morphogeny, the purpose of which is the development of life-forms or pyknons which are to appear on the various planets, stars, worlds and suns of the universe. It embraces the whole span of the life-aspect on the evolutionary side of manifestation. In this aspect is included also every conceivable adaptation of the universal principle of life from the beginning of its movement to the end. The universe is now functioning in and progressing through the vicissitudes of this stage. That is, all the present observable adaptations which the life-pyknon or principle is making, has made or will make, are embraced within the scope of what is here designated as the morphogenetic aspect of evolution. (*b*) The second and last division of the arc of evolution is called *Kathekos*, thus symbolizing the syncretism of the trinitarian aspects of kosmogenesis, *Chaos-Theos-Kosmos*, perfected and united as a result of the labors of manifestation. In this final summation of the labors of the life-wave as it has progressed from involution through all the devious manifestations of evolution are embodied the perfection and ultimate elaboration and expression of all the pyknotic tendencies which were established during the entire scope of space-genesis.

Thus it will be seen that the first three stages of space-genesis, called chaogeny, encompass the first three pyknotic processes or are analogous thereto while the latter called chaomorphogeny, the organization and ensoulment of space-forms, embraces the latter four, quartopyknosis, quintopyknosis, sextopyknosis and septopyknosis. This division obtains on the involutionary side of the great life-cycle. The upward arc of the Great Kosmic Egg or Cycle is also divided into two great stages, namely, Morphogeny (manifestation of life through the various forms which it assumes) and *Kathekos*, or the kathekotic plane of perfected triunity which is represented by the evolutionary union of *Chaos-Theos-Kosmos*. *Kathekos* would, therefore, symbolize the ultimate elaboration of Chaos into a well-ordered kosmos wherein are expressed all the possibilities which inhered in the archetypal plan of the Creative Logos or *Theos* and in which all had reached the ultimate perfection in the body of being of the Logos Himself. But the kathekotic plane is to be distinguished from the original *Chaos-Theos-Kosmos* represented as functionating upon the plane of non-manifestation during chaogeny. *Kathekos* symbolizes the perfected manifestations of the triune aspects of the Creative Logos through the perfected forms resulting from the labors of kosmic evolution, while *Chaos-Theos-Kosmos* symbolized, as a triune *glyph*, the Unmanifest Trinity in the primordial beginnings of space-genesis. One is the seed; the other the fully matured plant; one the egg; while the other is the full grown bird; one the root; the other the fruitage; one Alpha, the other Omega; one the beginning, the other the end. The end, however, is reached only that, in due time, the entire scheme may be commenced again, once more utilizing the results of the preceding scheme of evolution as the basis of the ensuing one. Thus after every Kosmic Day, commences the Kosmic Night. The succession of kosmic days and nights is infinite. This infinity of *becomings* in the life of the kosmos is a necessary outcome of eternal duration.

The above, thus briefly set down, is the symbolism of space-genesis. It is commended to the reader as a basis for the conception that the real, essential, perceptual space is something far more wonderful, more fundamental than either the geometrician or the metageometrician has ever dreamed of, and yet the latter's consciousness is undoubtedly being appulsed by the fingers of a new species of conceptualizations which, one day in the not too distant future, will arouse in it the faint hungerings after the realization of the real space-nature. These mathetic appetites thus brought into being will finally lead the human mind into the Elysian fields of kosmic consciousness where for another million years, perhaps, it may feed upon the mysteries and hypermysteries to be found in the granaries of the Space-Mind.

The study of space in its wider and deeper meanings is necessary in order that a clearer understanding of its true significance, as the subject of geometric researches, may be gained. It is confessed, however, that there is neither direct evidence nor implicative authority for any assumption that the view herein outlined affords any justification for the notion of the n-dimensionality of space. For, although the line of reasoning indulged in must lead inevitably to the conclusion that the worlds of spatiality, materiality, intellectuality and spirituality, essentially and fundamentally one so far as origins and qualities are concerned, were alike engendered by the same generating element, life; and that spatiality being the primal basis of the others is, nevertheless, under the exigencies of this aspect of the kosmos, highly susceptible to the mensurative requirements of the grossest, there appears to be no necessity for calling upon extraneous considerations for assistance in our efforts to comprehend the various connotations of the symbolism. Then, too, it is easily conceivable that under conditions where these elements, spatiality, intellectuality and materiality, are not only co-extensive but interpenetrative, there is no justification for the assumption that they must exist in layers or manifoldnesses or in discrete degrees, separated from one another as if they were constituted of different substances and occupied different spheres. For every single point in perceptual space is a focus for lines drawn through every conceivable grade of materiality, spatiality or intellectuality in the kosmos. And the same system of coördinates which is necessary and sufficient for the localization of a point in our space is also sufficient for the location of a point anywhere in the entire world of spatiality, intellectuality or spirituality. In fact, the external, visible worlds of materiality and spatiality are nothing more than the *mass-termini* of lines extending from divinity to physicality; from primordial originality to kosmic modernity and it is intellectually conceivable that progression back over the grooves made by these mass-termini of lines would lead directly and unerringly to originality itself. In spite of the manifold pyknoses which we have shown to characterize the symbolism of space-genesis it is a very simple matter; for the entire scheme could and must have proceeded along strictly tridimensional lines. Tridimensionality must have inhered in the primeval archetype of space or else it could not appear as an outstanding fact of perceptual space now; for all that we can now observe in space as characteristics must have first been included, enfolded, involved, before it could have been evolved. Hence, it is to be remembered that we are to-day dealing with the expressions of tendencies and principles which inhered in the manifested universe as potentialities in the very beginning.

The alphabet of space-genesis consists of five characters, namely, the point, the line, the triangle, the square and the circle. These are the pentagrammaton of space, of intellectuality, materiality and of spirituality. They constitute the basis of kosmic geometrism. With these all geometrical figures may be constructed; with them all magnitudes may be delineated and projected. They describe every conceivable activity of the Creative Logos and designate the bounds of the entire scope of motility of kosmogenesis.

In figure 19, are shown the dot, the line, triangle, square and the circle which together form the kosmic pentoglyph. The point symbolizes kosmic inertia, inactivity or the beginning of motion; the line is the first aspect of motion, the beginning of creation; the triple aspect of kosmogenesis is symbolized by the triangle, *Chaos-Theos-Kosmos*, the Unmanifest Trinity; the square emblematizes kosmic being in evolution; while the circle is the syncretism of all these and stands for the perfected kosmos, or the kosmos in process of perfection. Very truly did PLATO remark: "God geometrizes"; for the *pentagrammaton*—the point-line-triangle-square-circle—is the deity's way of manifesting Himself. But there is here no need for space-curvature nor for triangles whose value is greater or less than 180 degrees; there is no need even for the mathematical fourth dimension.

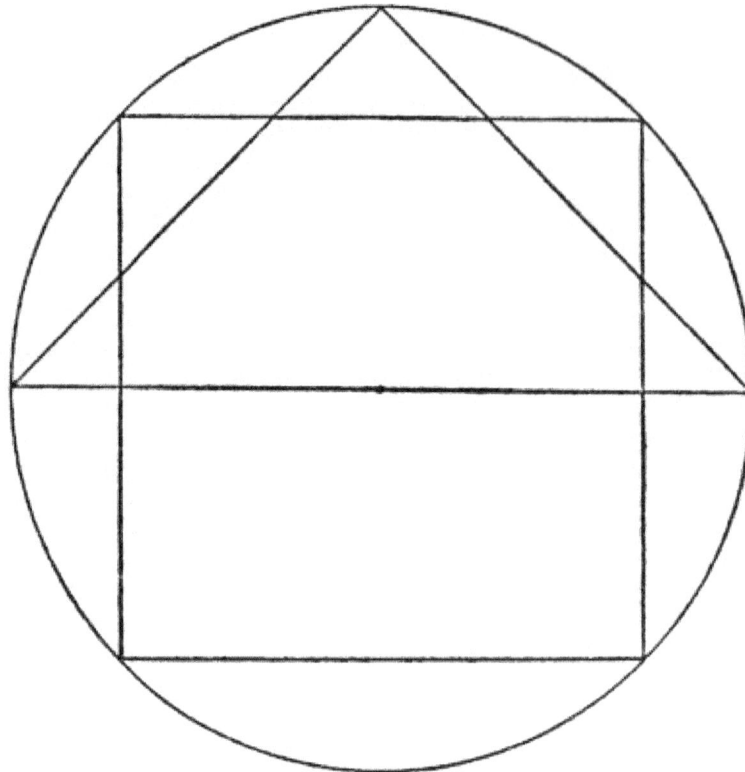

FIG. 19.—Kosmic Pentoglyph

It cannot be believed, however, that metageometricians are really in earnest in what they suggest of hyperspace and *n*-dimensionality; it cannot be believed that they are entirely satisfied with what they have found of the so-called hyperspatial, and yet, some of them are fanatically patriotic over the new-found domain; some are even intolerant. But there are others who look upon the fabric of metageometry as a stepping stone to space-realities, a mile-post on the path to the realization of a higher consciousness, the consciousness of the space-mind or kosmic consciousness. And may this not, after all, be the goal of the human intellect, now slightly distraught by the exuberances of youth and the joys of a new mental freedom? The work of the future mathematicians will be the destruction of the tumorous inconsistences to be found in the various non-Euclidean systems of geometrical thought, the elimination of the novelties and the nonsensicals, the synthesizing of those elements which are sanctioned by the space-mind and the building thereon a sane interpretation of space-phenomena in the light of illuminations received from greatly extended faculties and a participation in that larger consciousness into which the human race seems slowly to be immerging.

The domain of hyperspace is but the fairy-land of mathesis, peopled with goblins, gnomes, kobolds, elves and fays which are the spaces, dimensions, propositions, ensembles and theorems of the metageometrician. But like the fairies and nature spirits of the unseen about us, they have their bases in the real, objective world of facts however difficult it may be to establish their direct connection with it. As the gay, invisible sprites of phantom-land represent intelligent natural forces at work in the furtherance of the evolution of forms, so the impalpable things of mathesis are emblems of kosmic forces at work in the upbuilding of structures of higher consciousness which shall be towers of vision for the human soul whence it may view the hill-crests of infinite knowledge and the low-lying plains of kosmic mysteries.

Finally, it has been noted that space is the very consistence of the kosmos; it is the life, the form and both the outer and the inner manifestation of the combined life and form; it is reality, also illusion; it is concrete, also ideal. We have noted also that mind is consubstantial with space and that space gives it its inner life and nature as well as nourishes its outer growth and development. In fact, we have seen that space and mind are *one* essentially and that they exist as aspects of the same thing, life. In whatever way, then, that the mind normally views space that is the natural way. All attempts to deviate from the natural way are, therefore, unsanctioned by the nature of things. So long as geometry remains true to the nature of mind and space so long will it be valid universally and possessed of kosmic necessity and invariance. It behaves most unseemly when it departs from its fealty to the nature of things *per se*.

Both the outlook of the mind upon the objective world as well as its inlook upon its own states of consciousness or the subjective world are tridimensional. Its growth is tridimensional, its nature is likewise tridimensional, and there is not even the slightest tendence either to perceive, conceive or perform in a four-dimensional manner, mathematically speaking. Trace out the biologic development of each mental faculty, from the mind of the moneron to the mind of the most highly developed man and it will be found that everywhere and always, without variation or exception, the nature of each of these has been to express itself tridimensionally and naturally. There is not even the slightest sign of so much as a germinal appetence for the four-space; it would, therefore, seem almost a prostitution of mental faculty to divert mental energy into the seemingly useless channel of present-day metageometrical researches; yet, it must be admitted that even though the end sought cannot be attained, the final results of the intellectual delvings into the dread homogeneity of kosmic origins and the consequent realization of the awesome coevalism of mind and space whence shall arise the recognition of the wondrous unitariness of all existences, will be that we shall come upon that thrice mysterious contrivance—the Heart of Divinity, the Kosmic Space-Center in which abide the roots of the *Great All* in a marvelously indescribable unity and infinite originality.

We may conclude, then, that hyperspatiality and all its appurtenances are but the toys of the childhood of humanity. But, as the years pass and the days of maturity come on apace, it, toy-like, will also be discarded. And the mind will seize then upon the seriousness of reality just as the matured youth responds to the stern realities of life and manhood responsibilities. But no one can say that the toys of childhood are wholly useless; no one can say that the joys which they bring are entirely fatuous and unreal nor shall we attempt to intimate that mathetic contrivances are without utility, without purpose and significance in the life of the growing mind of humanity. But they, too, will pass away.

CHAPTER VIII

THE MYSTERY OF SPACE

The Thinker and the Ego—Increscent Automatism of the Intellect—
The Egopsyche and the Omnipsyche—Kosmic Order or Geometrism
—Life as Engendering Element—The Mystery of Space Stated—
Kathekos and Kathekotic Consciousness—Function of the Ideal—The
Path of Search for an Understanding of the Nature and Extent of Space
Must Proceed in an Inverse Direction.

The fragmentariness of the Thinker's outlook upon the universe of spatiality
is due to the inhibitive action set up by the constrictive bonds which his
complicate mechanism of intellectuality interposes between himself and
reality. The Thinker, who stands back of and uses the various media of
objective consciousness, such as the neural mechanisms, brain, emotions,
his individualized life-force and the mind which together make up the
instruments with which he contacts the sensuous domain, by adapting his
consciousness to these means, as the artisan utilizes his tools, constitutes his
own intellectuality. The intellectuality, then, is the totality of media by
which consciousness effects its entrance into the sensuous world and by
which it receives impressions therefrom. In other words, it is the sum of all
those qualities, operations, processes and mechanisms which are recognized
as constituting the *modus vivendi* of man's intellectuality, and these are, in
reality, nothing more than the ego himself.

Many have been inclined to regard that which has been called the ego as the
highest sovereign power in the state of manhood. He has been looked upon
as the final consideration in the constitution of the human being. But the
ego is an evolutionary product and the concomitant of self-consciousness
which is the I-making faculty in man's psychic life. It is that quality of
consciousness which makes man conceive of himself as a separate,

detached and independent being. It is a purely intellectual or tuitional product, and, as such, is to be differentiated from the intuitional or life-quality which is the essence of man's real selfhood. With respect to the Thinker, the ego occupies precisely the same status as the agent to his principal. As the agent is the representative of the principal in all matters which come within the scope of his prescribed jurisdiction so is the ego the agent of the Thinker who is a spiritual intelligence. Accordingly, from an ethical viewpoint the Thinker is responsible for the acts of the agent and can in no wise escape the penalties accruing as a result of the agent's violations. Just as a commercial firm sends out a representative for the collection of data concerning certain phases of its business or it may be of any business or the entire world market so the Thinker projects his own consciousness into the mechanisms which are in their totality the egoic life. That is, he sends out his agent, the ego, into life and into the objective world of facts and demands that he shall convey to him, from all points of the territory which he is expected to cover, reports of his findings. Of course, these reports which are transmitted by the ego (the intellectual mechanism of the Thinker) are more or less well prepared summations of his individual observations and deductions. These are the percepts which the ego presents to the Thinker's consciousness. Concepts are formed by the Thinker in his treatment of these sense-presentations. It very frequently happens that the ego transmits reports which, for one reason or another, give very imperfect knowledge of the matter which his reports are designed to cover. Often it is necessary that additional and supplemental reports be made about the same thing, and even then, it is well-nigh impossible, if not quite so, for him fully to cover every detail of the matter under consideration and in no case is it possible for him to do more than report on the superficialities of the question under scrutiny. If the ego, in his operations, be imagined to be hampered by similar circumstances and difficulties as those which would ordinarily beset a commercial attaché it will then be clear that his reports must ever be fragmentary because of the inaccessibility of much of the data which would be necessary for a full report, and further, because of the inadequacy of his methods and means of gathering data due to the inherent limitations of his capabilities, endurance and perspicacity and innumerable other limitations and difficulties which must be faced in all search for the real. So that, while the sufficiency of the means which the ego enjoys at this stage for all practical purposes is granted no hesitancy is entertained when it comes to a discovery of the reals of knowledge in declaring their insufficiency.

Then, too, when it is remembered that these egoic reports are in the nature of neurographical communications which are similar to telegraphic despatches and must pass through several stations, as ganglia, etc., often being relayed from one to another, it will be quite apparent that much, even of the original quality of the missives forwarded, will have been lost or radically changed in some way before it is finally delivered for the inspection of the Thinker himself. It not infrequently happens, even in perfectly normal beings, that the ego in filing, recording, transcribing, interpreting, translating and otherwise preparing these data for the Thinker's use, lets a cog slip, misplaces some of the data, loses or destroys fragments of it and so is unable to maintain a complete portfolio of his materials.

As the Thinker is entirely dependent upon his agent, the ego, for the trustworthiness of his information covering the matter of the sensuous world it is obvious that at best his information is very fragmentary indeed, and necessarily so when it is considered that the *modus operandi* of his agent and the difficulty of his operations are so complicate as to magnify the obstructions in the way to complete freedom in this regard.

To continue the similitude of principal and agent it may be asserted that it is also true that the commercial house that sends out its attaches frequently will send letters containing directions as to procedure, sometimes censuring for past delinquencies and sometimes commending for praiseworthy deeds; and this, too, in addition to the original instructions which were given at the outset. It even comes to pass that the home office, because of some meretricious accomplishment, as the marked increase of efficiency shown by the agent's close application to his duties and the consequent success of his operations, confers certain favors upon the agent or removes some of the restrictions which were originally imposed, gives an increase in salary or promotes the agent to a higher and more lucrative office with larger powers and greater authority. This is analogous to what the Thinker does for the ego. For he not only receives reports from the ego, but often, in the shape of intuitions, gives additional information as to the proper manner of doing things, sheds more light upon some obscure operation, commends for duty well performed, condemns for failures or for wrong-doing, rewards arduous toil with greater powers of vision, keener insight, greater capabilities; in fact, promotes the ego in the sight of other egos by marking him out as an exceptional ego. But the curious aspect of this procedure is that, in time and after the ego has been repeatedly commended and promoted and otherwise favored by the Thinker, he begins to think that he owns the firm, that he is the life and main support of the whole corporation. He becomes arrogant, self-willed and finally falls into the illusion that he alone is responsible for the phenomenal success of the firm. This is the source of that illusion of the intellect which makes itself think that it, the ego, is all there is to man, that his instruments of operation in the objective world are the only kind of instruments that may be used; that his method of gathering data about things is the only safe and sure method; and so it develops that the intellectuality is the source of man's separateness, his individuality and his apparent aloofness from other men and things. It is, of course, needless to point out that in this way the intellect comes to be the tyrant of man, ruling with a rigid monopoly and as an all-exclusive autocracy.

From the above implications it would appear that the intellect and the intuitive faculty are two separate and distinct processes, and so they are. One is the inverse of the other. The tendence of the egoic life or the intellect is for the external while the intuition is an internal process. The intellect acts from without towards the interior while the intuition acts from within outward. The intellect is the product of the intuition which is another term for the consciousness of the Thinker on his own plane. Just as the child lives a separate and distinct, though dependent, life from the parents so the intellect has a *modus vivendi* which is distinct and separate from that of the Thinker, and yet it is in all points dependent upon the life of the Thinker. Here again, we find an analogy in the relation of the child to the parent. As some children are more amenable to the will of the parent than others, so, in some persons, the intellect is more amenable to the action of the intuition than in others. Yet it is a certain fact that the more the outward life is governed by the intuition, i.e., the more the intellect responds to the intuitive faculty of the Thinker, the higher the order of the life of the ego and the more accurate his decisions and judgments. In fact, it assuredly may be asserted that the place of every individual in the scale of evolution is determined in a very large measure by the degree of agreement between the intuition and the intellect or by the ease with which the intuition may operate through the intellect as a medium. At least, the quality of one's life may be determined directly by these considerations.

The Thinker being himself a pure spiritual intelligence, living upon the plane of spirit and therefore unhampered by the difficulties which the ego meets in his operations in the objective sensorium, and possessed of far greater knowledge, is correspondingly free from the limitations of the ego and very naturally closer to kosmic realities. Hence, he is better situated for the procurement of correct notions of relations, essentialities and the like. It is believed, therefore, that in the proportion that these two processes, the intellectual and the intuitional, are brought, in the course of evolution, to a closer and more rigid agreement, in the proportion that the Thinker is able to transmit the intuitograms in the shape of concepts or that the intuition is made more and more conceptual, in just that proportion is humanity becoming perfect and its evolution complete. The difficulty found to inhere in the conceptualization of intuitions so that they may be propagated from man to man seems not to lie in the Thinker himself, but more essentially in the ego, in the intellectuality and its complicate schematism or plan of action. It would appear, therefore, that the only way of escaping or transcending this difficulty is for the ego so to refine his vehicles or so facilitate his plan of action by eliminating the numerous relays or sub-stations intervening between the consciousness of the Thinker and that which may be said to be his own that the transmission of intuitograms may be accomplished with the greatest ease and clearness. While no attempt will be made to indicate the probable line of action which the ego or objective man will adopt for this purpose, it is believed that it may be said without pedanticism that the only true method of attaining unto this much desired state of things is, first of all, by assuming a sympathetic attitude not only towards the question of the intuition itself but to all phenomena which are an outgrowth of, or incident to, the manifestations of the intuitive faculty through the intellectuality, and second, by the practice of prolonged abstract thought, this latter procedure effecting a suspension of the intellectuality temporarily at the same time allowing it to experience an undisturbed contact with the intuitional consciousness, thereby laying the basis for future recognition of its nature and quality. It would seem that these two conditions are absolutely necessary in order that a more congruent relationship may be promoted between these two cognitive faculties.

Ordinarily, it would appear that the philosopher who is undoubtedly inured to the necessities of continuous abstraction or the mathematician whose most common tasks naturally fall in this category would be among all men most apt to develop to the point of conceptualizing intuitograms readily, yet it seems that this is not the case. And there is good reason for it. The mind of the philosopher and the mathematician is intellectual rather than intuitional and is, therefore, wedded to matter, to the action and reaction of matter against matter and hence operating in a direction at variance with the trend of an intuitional mind. And this condition is undoubtedly due to a lack of a sympathetic attitude towards this species of consciousness. At any rate, it is thought that a too great anxiety in this respect need not be entertained by humanity at all, for the reason that in the case of a faculty, the rudimentary outcroppings of which are so marked and universally observable and existing in greater or lesser degrees in various human beings, there is ample evidence for the belief that it is being carefully and duly promoted by a well-directed evolution of psychic faculties and powers, so that at the proper time, determinable by the state of perfection reached by the intellectuality or the ego in the operation of his cognitive processes, the much desired agreement of these two faculties will have been realized and the conceptualization of intuitograms into propagable conceptions an accomplished fact. Until this goal shall have been reached and the intuition shall have overshadowed the intellect as the intellect now overshadows the intuition; or the consciousness of the ego, derived from the interplay of the Thinker's consciousness among the various elements which constitute the ego himself, shall have been merged with that of the Thinker, the outlook must remain fragmentary, only becoming a well-ordered whole as the barriers of dissidence are broken down in succession.

The evolution of consciousness, from the simple, undifferentiated *moneron* to the differentiated cell and from that to the cell-colony and from the cell-colony to the organism, traversing in successive paces through all the stages of lower life—mineral, vegetable and animal—to the stages of the simple, communal consciousness of the higher animals, to the self or individual consciousness of the human being, each requiring millions of years for its perfection before a more advanced stage is entered, has been one continuous relinquishment of the lower and less complicate for the higher and more complex expression of itself through the given media. When a newer and higher stage of consciousness is being entered by humanity its appearance or manifestation is first made in the most advanced of the race and that only in a dim, vague way. This rudimentary condition persists for some time, perhaps many thousands of years, then the faculty becomes more general in appearance, the number of advanced individuals increases, and consequently, as in the case of the intuitive faculty, it becomes universally prevalent in all humanity; becomes transmissible as so-called "acquired characters," and then appears as the normal faculty of the entire human family cropping out in each individual. Thus, in passing from the few advanced ones in the beginning to that stage where it becomes the common possession of all, a faculty requires many thousands of years for its perfection, and especially has this been true in the past history of the development of human faculties. But it is believed that the sweep of the life current as it proceeds from form to form, from faculty to faculty, gains in momentum as it proceeds, so that in these latter years due to the already highly developed vehicular mechanisms at its disposal not so great a period of time as formerly is required for the *out-bringing* of a new faculty. It might well be that while in the past hundreds of thousands of years were necessary in the perfection of organs and faculties, in these latter days only a few thousand, perhaps hundreds, may be necessary and that in the days of the future not even so many years may be required to universalize a faculty. And especially does this appear to be true in a state of affairs where so large a number of persons are beginning consciously to take their evolution in hand and by volitional activities are supplying greatly increased impetus to their psychic processes which under ordinary, natural methods would be considerably slower in their development. It is quite obvious that all cultural efforts when applied to the betterment of a given plant, animal or faculty result in a corresponding hastening of the process of growth far in excess of what that growth would be under normal, natural conditions. All the present faculties possessed by man are remarkably susceptible to cultural influences; in fact, the standing edict of ethical and social law is that the human faculties must be cultivated as highly as possible, thereby giving the spirit a more perfect medium of expression. These observations, therefore, lead irresistibly and unavoidably to the conclusion that the time for the upspringing of the intuitional faculty in the human organism is even now upon us, that undoubtedly in certain very advanced ones it has already reached a notable degree of perfection and is rather more general than would appear in the absence of careful investigation.

Now, just as the intellect has made for individuality, has emphasized the separateness of the Thinker's existence from that of other thinkers, has developed self-consciousness to a very high degree, even pushing it far over into the domain of the higher consciousness to the temporary obscuration of the latter, so the intuitional will make for union, for the brotherhood of man, for co-operation and for the common weal. Through it man will come gradually into the consciousness that fundamentally, in his inner nature, in every respect of vital concern, he is at-one with his fellowmen and not only with the apparent units of life but with all life as expressed in whatsoever form throughout the universe. Then, too, he will be closer to the reality of things, of actions and natural processes; in fine, he will have begun the development of the space-mind which will bring him to the knowledge that he is one with space also and, therefore, with the divine life of the world.

One of the peculiarities of the vital force which shows itself in the consciousness as man's intellect, is its growing *automatism*, or that tendency which enables the consciousness to perform its functions automatically and thus allow opportunity for the development of newer and higher faculties. Actions, oft repeated, tend to become automatic. This is also true of thought and consciousness. It is one of the beneficent results of abstract thought that it develops, or tends to develop, a kind of automatism whereby a marked saving in time and energy is effected. This affords opportunity for other things. It is undoubtedly true that in the days of the truly primitive man his consciousness was more completely engaged in the execution of the ergonic functions of cells, organs and tissues; that all those processes which are now said to be involuntary and reflexive were at one time, in the distant past of man's evolution, the results of conscious volitions. This is a condition which must have preceded even the development of the intellect itself. Indeed, there could be no intellect in a state where the entire modicum of consciousness was being utilized in the performance of cellular and histologic functions.

The rise of the intellect must have been in direct ratio to the development of automatism among the cells, tissues and organs, so that as these came gradually to perform their special labors reflexively the intellect began to be formulated and to grow, at first only incipiently, then more and more completely until it reached its present state. At the present stage of its evolution, a great deal of the labor of the intellect is beginning to fall into a kind of increscent automatism, although only rudimentarily, in many instances. Yet, as a result of this tendency, quite the whole of the phenomena of perception is characterized by a sort of automatic action. And the mind perceives without conscious volition. Many of the steps of conceptualization are automatic, in part, if not wholly. Certain it is that impulses once set in operation whether consciously or unconsciously continue to act along the same line until exhausted or until the end has been attained. Consequently, it is a proven fact that often serious mathematical and philosophic problems have been solved by the mind long after any conscious effort to solve them had ceased. Often solutions have been arrived at during sleep. Many such cases might be cited, but the phenomenon is now so common that almost every one can cite some experience in his own life that will substantiate the claim.

There is no doubt but that these phenomena are evidences of a reflexive development in the intellect. The time will come undoubtedly, and necessarily so if the intellect is to give way to a higher faculty, which shall be as much above the intellect in its grasp of things as the intellect is now above the simple consciousness of the lower animal, when quite the entirety of our intellectual processes will become automatic or self-performing. What then remains of the egoic schematism, after its transmutation or elevation as the organ of the intuitional consciousness will be utilized as the organ of the Thinker's involuntary cognitive processes. This will mean that all of that laborious ideation which is now the abstract thought of the Thinker will be performed automatically, leaving the higher aspect of the egoic consciousness free to conceptualize or intuitograph the intuitions. Perceptualization then will be replaced by conceptualization. This latter will occupy about the same status as the former does now. And necessarily, perception will become more complex. In other words, while we now perceive simple percepts which are again arranged into concepts making a composite picture of the object, we shall then be taking in the composite picture of the object at first hand, thereby dispensing with the rather slow process of perception as it now operates. We shall still be perceiving, but what we perceive will be concepts rather than percepts, as at present.

The increased powers of intellection gained as a result of the increscent *automatism* in the intellect, the flowering forth of the intuitive faculty and the general enhancement of the intellect throughout all its processes will enable it to entertain concepts or composite pictures of things just as readily and as perfectly as it can at present deal with a single percept. Concepts will be replaced by super-concepts or intuitographs. Increased perspicacity will enable the Thinker to manipulate the concepts and intuitographs with the same ease and readiness and withal the mind will have attained unto an almost unrealizable freedom in its search after truth.

The outcome of this new adjustment which, of course, will not spring up at once, but by insensible degrees, will be the clarification and unification of our knowledge. It will mean also the simplification of it; the obviation of diversities of opinions, the springing up of a new and winnowed system of philosophy which shall be the true one; further, it will imply the lessening of the probability of error in our judgments and conclusions; the removal of illusion to a much larger degree than to-day is possible and the realization by every one of something of the essence of things, of causes and effects, of actions, operations, natural forces and laws; in fact, a condition of mind which will present to the consciousness the simple truth above every conceivable phase of kosmic life which may come within the scope of the Thinker's observation.

The further implications of this view are that there is a difference between the Thinker and the intellectuality. The Thinker is eternal and partakes, therefore, of the very essence of primordial originality while the mentality is an artificial process, the resultant of the adaptation of the Thinker's consciousness to his vehicular contrivances of objective cognition and the interplay of his life among them.

If the appearance of a choppy sea disturbed by the passage of a brisk breeze over its surface be imagined, a similitude of the great ocean of life may be envisaged. The wavelet crests symbolize the egos; the base of the wavelet which is one with the great sea of water represents the Thinker which is one with the divine life and consciousness of the kosmos. Just as wavelet crests are continually springing up and falling back into the sea, so are egos continually being cast forth and reabsorbed into the universality of life only to be recast, as a wavelet crest or ego, upon the surface of the moving ocean of life. *And so, in this respect, the universum of life and consciousness which are essentially one is in a constant state of ever-becoming, un-becoming and re-becoming.*

Another implication is that, on account of the diversity and complexity of the means of contact with the external world, it is not possible for the ego to arrive at more than a fragmentary understanding of even the latent geometrism of life, mind and materiality. In our examination of the sensuous world, we are very much like the three blind men set to examining an elephant. One set to scrutinizing his trunk by means of his sense of feeling. When asked for his judgment as to what the elephant was he declared it was a snake; a second who began with the legs found it to be like huge pillars; and a third who caught hold of the elephant's tail and declared the elephant to be like a rope. Each one of the blind men described what he was able to perceive. To each what he felt was all there was upon which he could render judgment. And so, artists, philosophers, mathematicians, musicians, mechanicians, religious seers, metaphysicians and all other types of mind, are just so many blind men set to the examination of an elephant, or the sensuous world. Each one confidently believes his view to be correct; each one is satisfied with the deliveries of his senses. Yet no one of them is wholly correct, no one of them has seen every phase and aspect of the problem. Does it not, therefore, appear to be the more reasonable and urgent that the view which synthesizes the judgments of all the possible examiners thereby constructing a composite idea of the entire mass of judgments is the more reliable and the more correct?

Referring again to the dual intelligence, the ego and the Thinker, which together constitute man, it is deemed necessary, in order to present the concept of this duality to the mind of the reader in the way that shall enable him easily to recall it, to designate the egoic intelligence as the *egopsyche*, and the Thinker's intelligence as the *omnipsyche*.

The egopsyche is the I-making faculty, the faculty of self-consciousness and the synthesis of all those psychic states and functions known as the intellect or mind and includes the ethical aspect of man's nature. The omnipsyche is the organism of kosmic consciousness, the space-mind, or man's higher self and that which connects with or allies him to all life; it is the basis of human unity and of unity with divinity, just as the egopsyche is the basis of separation and individuality; it is the organ of direct and instantaneous cognition and the permanent essence which has persisted through every form which the being, man, has ever assumed and through every stage of human evolution. In it are stored up the memories of the Thinker's past, the secrets of life, mind, being, reality, and the history of life from the beginning; in it also the plan of action for the future of the life-wave as it passes from plane to plane, from stage to stage, and from form to form. It is the spark from the flame that is never quite free from its source; it is the continuous spark, the prolonged ray which does not go out and cannot be extinguished. It is that in man which when full union therewith has been attained makes him a god in full consciousness.

The omnipsyche is really a neglected and overlooked factor in the doctrine of evolution. Evolutionists, while they claim life to be continuous and that man has come through all the kingdoms of nature in succession and has spent millions of years in the perfection of his various organs, faculties and stages of consciousness, make no ample allowance for what is in reality the basal element in evolution—a continuous, persisting, permanent life-force which does not lose its identity from the beginning to the end of the process. This fact—that that spark of life which set out upon the evolutionary journey as a moneron has glowed steadily from that stage to manhood, maintaining meantime its original purposiveness and intent— seems to be the most obvious consideration of the whole doctrine, yet it has been more or less completely ignored. The elementary requirements of evolution would seem to establish clearly the necessity for some such eternally persisting principle as the omnipsyche which is capable of such subtle adaptations to every conceivable form of life and in which should be gathered up the evolutionary results of every life-cycle. For this purpose the omnipsyche or unifying principle in man was designed from the beginning and it is that which constitutes the basis of his intellectual nature while in a far larger sense it is the divinity in man himself. It is indeed strange that so important a factor as the omnipsyche should have been omitted by evolutionists. Yet it can be accounted for upon the grounds of the purely mechanistic character of all intellectual attempts at solving the problems of vital manifestations. But so long as men rely upon mechanical explanations of such phenomena so long will they be prone to overlook the very essentialities of the problems which they devoutly wish to solve. The continuity of the physical germ-plasm of the human species,[27] now quite generally admitted, would suggest, it seems, an analogous condition to the continuity of the psychic plasm called the omnipsyche, the only difference being that the omnipsyche is an intelligent factor while the physical plasm is a medium of transmission though non-intelligent. The omnipsyche is, therefore, the psychic reservoir of evolution into which are stored the transmuted psychics of moneron, amoeba, jellyfish and every other form which it has ensouled and acts as the storeroom of man's psychic operations as well as the source of his intellectuality.

We turn now from the study of a sketch of the mechanism of man's consciousness which gives at its best only a fragmentary view of the universe of spatiality to a consideration of space itself in the light of its interrelational bearings upon the question of intellectuality.

In the chapter on the "Genesis and Nature of Space" we have, in tracing out the engenderment of space, proved it to be basically one with matter (and indeed the progenitor of matter), also with life and consciousness. Further, it has been shown that all the characteristics of materiality are due to the adaptation of consciousness to it and that out of this adaptation grew the intellectuality. A close approximation to this view was maintained by KANT when he discovered that our faculty of thinking or the intellect only finds again in matter the mathematical order or properties which our faculty of perceiving or consciousness has deposed there. It appears, therefore, that when the intellect approaches matter or spatiality it finds always a ready yieldance to its demands simply because intellectuality has previously established therein the delineation or map of the path over which it necessarily must traverse in its examination of the object of its pursuit. In other words, the kosmic mind in engendering materiality and spatiality has set up therein a kosmic order or geometrism. Both motor and intellectual progress, therefore, can be made through the world of spatiality because of the immanence of this kosmic geometrism which lies latent in the very fabric of the world of substance fashioning both the character and the nature of the intellect as well as of space itself. So that there is a perfect congruity subsisting between spatiality and intellectuality. Accordingly it is impossible for either one or the other to transcend the grim grasp of the mathematical order which binds them in such lasting and fundamental agreement. Extra-spatiality may degrade itself into spatiality, and indeed in the very nature of the case, does so degrade itself, yet spatiality can never raise itself beyond the limits set by its engendering parent. Materiality may become more and more spatialized and consciousness more and more intellectualized, but they must proceed hand-in-hand one not superseding the other.

Being the essence of the natural geometry which is everywhere immanent in the universum of matter, space becomes an organized and ordered extension, in fact is the totality of such organized and ordered extension, which conforms to the latent geometrism the engenderment of which it is the sole cause in the last analysis. Does it not appear then that all that mass of artificial geometry which has sprung up as a result of departures made from this natural geometry is utterly baseless and most certainly lacking in the kosmic agreement which spatiality lends to our primary conceptions? Of course, it is admittedly possible to devise certain conventional forms of logic and endow them with all the evidences of a rigid consistency but which, because of their purely artificial character, will fall far short of any real conformity to the potential geometrism which has been established in spatiality. And this fact is of utmost significance for all those who seek to find justification either logically or naturally for the existence of a multi-dimensional quality in space; for, if a clear, discriminative conception as to the categorical relationship, each to each, of the two kinds of geometry be carried in mind, it will not be easy to confound them neither will it be difficult to discern where the one ends and the other begins.

Now, the fourth dimension and the entirety of those mathematical speculations touching upon the question of hyperspace, dimensionality, space-curvature and the manifoldness of space are purely conventional and arbitrary contrivances and do not meet with any agreement or authority in the native geometrism which we find inhering in space and which the intellect recognizes there. This conclusion seems to be obvious for the reason that, in the first place, the non-Euclidean geometries have been constructed upon the basis of a negation of the latent geometrism of space and intellectuality; and if so, is it reasonable to expect that either they or any of their conclusions should accord with the nature of that form of geometry so admirably delineated by EUCLID? Obviously not. It is a matter of historical knowledge that the whole of the artificial non-Euclidean geometries consists of those purely conventional results which investigators arrived at when they denied or controverted the norms supplied by the natural geometry. When metageometricians found that they could neither prove nor disprove the Euclidean parallel-postulate they then set upon the examination of idealized constructions which negatived the postulate. The results, thus obtained, although self-consistent enough, were compiled into systems of geometry which naturally were at variance with each other and with this inherent geometrism which is found in spatiality and answered to by the intellect both normally and logically.

Furthermore, there is another consideration which to us seems to be equally if not more forbidding, in its objections to the coördination of the two systems of geometry, and that is the fact that the geometry of hyperspace is denied the corroborative testimony of experience and this is true of practically the whole of its data. Indeed, there is perhaps no single element in its entire constitution which claims the authority of experience. This is undoubtedly the weakest point in the structure of the hyperspatial geometries. Contrarily, such is not the case with the natural geometry; for, in this, the intellect in retracing its steps over the path laid out by that movement which has at the same time created both the intellect and spatiality, finds an orderly and commodious arrangement into which it naturally and easily falls. So exact is this agreement of the intellect with the kosmic order that if it were possible to remove the whole of spatiality and materiality there would still be left the frame work which is this latent geometrism of kosmogenesis. But the fact that the intellect naturally fills all the interstices of materiality and spatiality, fitting snugly into all of them as if molded for just that purpose, by no means warrants the assumption that it would or does also fit the engendering factor which has created these interstices. The frame work, the order or the geometrism of the kosmos has been established by life acting consciously upon the universum of materiality. And in order to establish this geometrism life had to be mobile, active, creative. It could not remain static, immobile, and accomplish it. Being mobile, dynamic, creative, it passes on. It is like a fashioning tool which the cabinet makers use in cutting out designs upon a piece of wood. It moves, and keeps moving until the design is finished, and then it is ready for more designing. Life is like that. It cuts out the designs in materiality, fashions the form, molds the material, and passes on to other forms. The intellect fits into these designs gracefully. But what it finds is not life itself, only the design which life has made. Hence, as there is neither an empirical spatiality nor materiality in conformity with which the artificial geometry of the analyst may be said to exist, and as it may not be said to conform to the path which life has made in passing through either of these, it is absurd to predicate it upon the same basis as the natural geometry. And so, we are forced, in the light of these considerations to deny the validity and hence the acceptability of the non-Euclidean geometries as either reasonable or warrantable substitutes for the Euclidean, and denying which we also formally ignore the claims of the fourth dimension, as mathematically designed, to any legitimate anchorage in either our vital or intellectual movements.

It has been shown that the flow of life, as it describes that movement which we call evolution, engenders simultaneously and consubstantially spatiality, materiality and intellectuality, and these, in turn, the natural order or geometrism everywhere immanent in the universe; and that automatically, one out of the other and each out of the all, these constitute the totality of kosmic fundamentals. Also we have sketched the mechanism of man's consciousness and discovered how, in its evolutionary development it has divided into two aspects, the egopsychic and the omnipsychic, and these two factors ally him definitely and adequately to the world of the senses and to the world of supersensuous cognitions. And thus we have cleared up some of the misconceptions which had to be confronted and made more easy the approach to the central idea, thereby conserving the substantiating influence which a general and more comprehensive view of the whole would naturally give.

The totality of kosmic order is space. It is circumscribed by an orderless envelope of chaos just as the germ of an egg is surrounded by the egg-plasm. The organized kosmos is the germ, kernel or central, nucleated mass, enduring in a state of becoming. Involutionary kathekos or primordial chaos is the egg-plasm which nourishes the germ or the kosmos and is that out of which the germ evolves. Kathekos or chaos is the unmanifest, unorganized, unconditioned, unlimited and undifferentiated plasm. Space is the manifest, limited, finite, organized germ that, feeding upon the enveloping chaos, exists in a perpetual state of alternate manifestation and non-manifestation—appearing, disappearing and reappearing indefinitely.

The appearance of the kosmos as an orderly elaboration of the involutionary phase of kosmogenesis, in so far as kosmic order may be said to be an accomplished fact, marked the turning point in that procedure whose function it was to make manifest a universe possessing certain definite characteristics of orderliness; but the kosmos, as it now stands, may not be thought of as having attained unto a state of ultimate orderliness. The idea meant to be conveyed is that between the point of becoming and the actually pyknosed, or solidified stage in the process of creation there is a more or less well defined line of demarkation cutting off that which is spatiality from that which is non-spatiality. Beyond the limits of spatiality is an absence of geometric order. Here geometry breaks down, becomes impotent, because it is an intellectual construction; at least, it is not so apparent as in the manifested kosmos. It is a state about which it is utterly futile to predicate anything; because no words can describe it. The most that may be said is that it is absence of geometric order as it inheres in space. And if so, all those movements comprehended under the general notions of spatiality, materiality, intellectuality and geometricity have both their extensive and detensive or inverse movements nullified in their approach to it. Involutionary *Kathekos*, therefore, may be said to be the primordial wilderness of disorder which outskirts the well laid-out and carefully planned garden of the spatial universe. We may excogitate upon some of the obvious functions of this kathekotic world-plasm; but in doing so we must leave off all attempts at a description of its appearance, its magnitude, extent or other qualities, and think only of its kosmic function. We cannot say that there is back of it a spatiality nor can we say that it is a spatiality; for whatever may be its extent or volume, it suffices that it may not be said to be space. It is chaos. Space is order, organization, geometricity. It cannot be said that there is a latent geometrism in chaos; because geometric order is found only in spatiality and is that which distinguishes spatiality from kathekosity or non-spatiality. Chaos is the lack of spatiality. This, of course, implies that it is impenetrable to the intellectuality or to vitality. All inverse movement such as is discovered as taking place in spatiality and which results in the phenomenalization of space runs aground when it strikes against the rock-bound coast of kathekosity. We can only say that it is both the point of origin for the evolving universe of life and form and its terminus. It is the nebulosity out of which the whole came and into which all is ultimately occluded.

A great and far-reaching error is made in all our thinking with respect to the kosmogonic processes when we postulate the complete absorption of chaos as an early act of kosmogony. Customarily, we think of kosmic chaos as a primordial condition whose existence was done away as soon as the universe came into active manifestation. This because it has been exceedingly difficult, if not quite impossible, for those whose privilege it was to determine the trend of philosophic thought to free themselves from the bondage of a dogma which owed its existence to a traditional or legendary interpretation of facts that ought never have been so interpreted. Chaos IS and EVER SHALL BE, so long as the universe itself lacks completion, fullness or perfection in purpose, extent and possibility. It is undoubtedly being diminished, however, in proportion as the kosmos is approaching absolute perfection. And when the last vestige of chaos disappears from the outerskirts of the maturing kosmos there shall appear a *glorified universe* of indescribable qualities.

Space being a perception *a priori* cannot be determined wholly by purely objective methods. The yard-stick, the telescope and the light-year are objects which belong exclusively to the phenomenal and with them alone never can we arrive at a true conception of the nature of space. We can no more demonstrate the nature of space by the use of objective instruments and movements than we can measure the spirit in a balance. Certainly, then, it cannot be hoped that by taking the measurement of space-distances in light-years, or studying the nature of material bodies, we shall be able to fathom this most objectively incomprehensible and ineluctable thing which we call space. It is such that every Thinker must, in his own inner consciousness, come into the realization of that awfully mysterious something which is the nature of space both as to existence and extent by his own subjective efforts unaided, uncharted and alone. When we measure, weigh, apportion and otherwise try to determine a thing we are dealing with the phenomenal which is no more the thing itself than a shadow is the object which casts it.

What does it matter that metageometricians shall be able to demonstrate that space exhibits itself to the senses in a four- or *n*-dimensional manner? Granting that they may be able to do this, if merely for the sake of the discussion, when they have finished, it will not be space that they have determined, but the phenomena of space, its arborescence, while space itself remain indeterminate and unapproachable by phenomenal methods. If there are curvature, manifoldness and *n*-dimensionality these are not properties of space, but of intellectuality in its cultured state and when it is, therefore, removed from the native state of conception. Scientists may be able to weigh the human body, count every cell, name and describe every nerve, muscle and fiber; they may even be able to know it in every conceivable part and from every physical angle and relationship, and yet know nothing of the life which vitalizes that body and makes it appear the phenomenal thing that it is. So it is not by instruments which man may devise that we shall be able to determine the true nature and purpose of space. We must adopt other methods and means and assume other angles of approach than the purely objective in order to comprehend space which, being the sole inherent aspect of consciousness, can be understood best by applying the measures which the latter provides for its understanding. It would appear, therefore, that the best study of space is the consciousness itself, knowing which we shall know space.

The universum of space, including the phenomenal universe, and its relation to consciousness may be likened to a conical funnel whose base represents the phenomenal world of the senses and whose apex or smallest point represents ultimate reality.

In Figure 20, we have endeavored to symbolize graphically this conception of space. The base marked "Sensorium" represents the sensible world. That marked "Realism" symbolizes the ultimate plane of reality, the inner essence of the world, the plane of "things-in-themselves."

The cone arising from the base "sensorium" symbolizes the objective world as compared with consciousness; the subverted cone, with apex in the sensorium, represents the evolving human consciousness.

The successive bases have the following symbology: Self-consciousness, Communal Consciousness, Mikrocosmic Consciousness, Makrokosmic or

Universal Consciousness, the Plane of the Space-Mind Consciousness, Divine Consciousness, Kathekotic Consciousness, or the Plane of Final Union with the Manifest Logos.

Self-consciousness is that form of consciousness which enables the ego to become aware of himself as distinguished from other selves or the Not-self; the Omnipsychic or Communal Consciousness is that form of consciousness from which arises the realization by the Thinker of his oneness with all other thinkers and with other forms of life. Mikrocosmic consciousness denotes a still higher form of consciousness, as that which enables the Thinker to become conscious of his living identity with the life of the world or the planet on which he lives. It represents a stage in the expansion of consciousness when it becomes one with the consciousness of the planet upon which it may

be functioning. Makrokosmic consciousness accomplishes the awareness of the Thinker's unity with the life of the kosmos or universe. The space-mind and the consciousness which constitutes it enable the Thinker to comprehend the originality and the terminality of kosmic processes. It is archetypal so far as the life-cycle of the universe is concerned because the beginning, the intermediate portion and the ending of the kosmos are encompassed within it. Divine consciousness is that form of consciousness which arises upon the unification of the Thinker's consciousness with that of the manifest deity; it is, in fact, omniscience. The kathekotic consciousness belongs to the ultimate plane of reality; to kosmic origins and chaogeny, and therefore, pertains to the plane of non-manifestation.

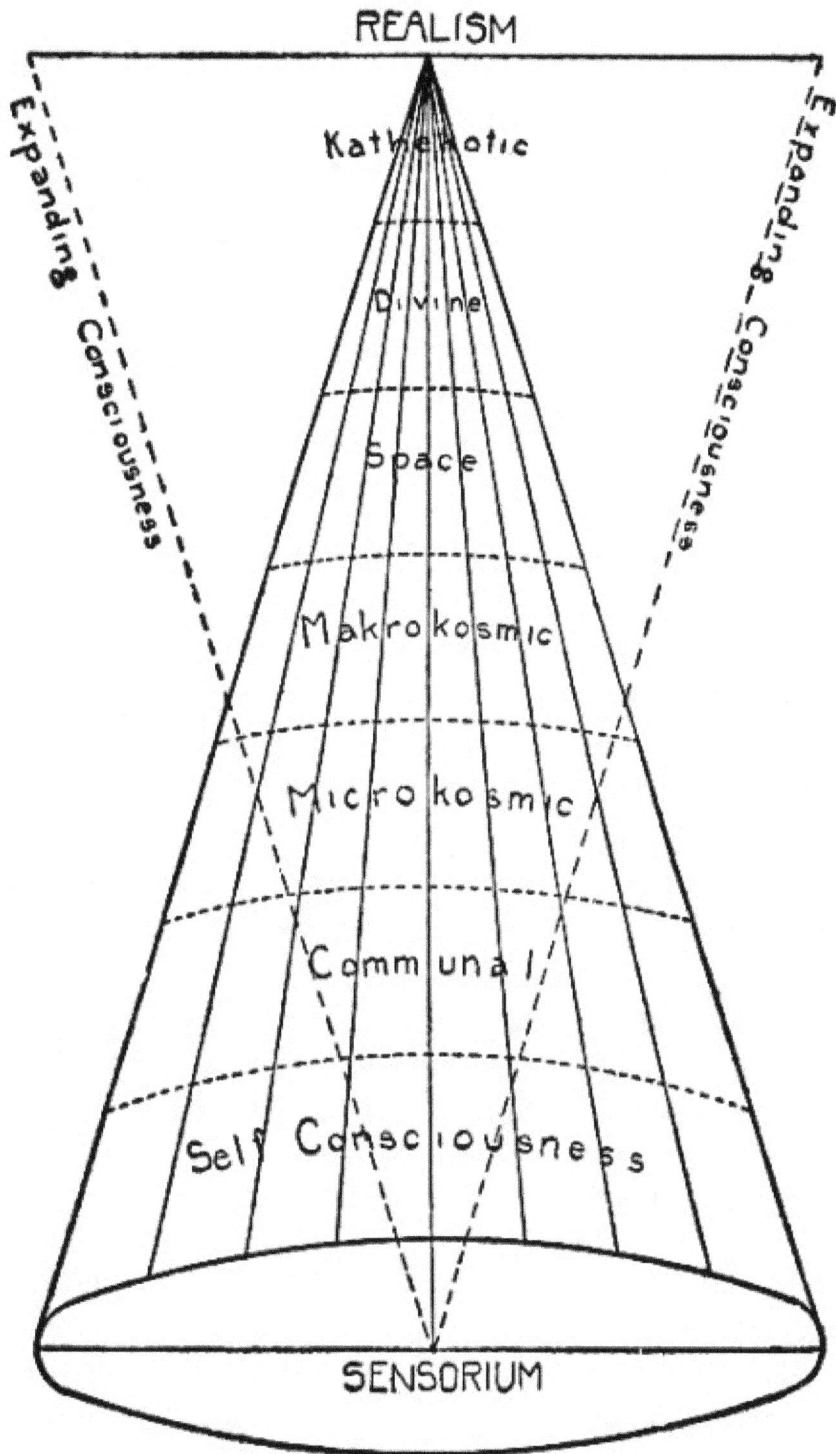

FIG. 20.—Kosmos and Conciousness

The implications are that in comparison with the sensorium, the Thinker's consciousness is a mere point in space. It is, in reality, so small and insignificant that the extensity of the physical world or universe seems unlimited, unfathomable in meaning and infinite in extent. But as his consciousness expands, as it passes, in evolutionary succession from one plane of reality to another and higher one, the illimitability, the incomprehensibility and infinity of the universe grow ever smaller and smaller, until the plane of divine consciousness is reached. Then the previously incomprehensible dwindles into insignificance, lost in the real illimitability, infinity and unfathomability of consciousness itself. Kosmic psychogenesis, as exhibited and specialized for the purposes of the evolution of the Thinker, can have no other destiny than the flowering forth as the *ne plus ultra* of manifestation which is nothing short of unification with the highest form of consciousness existent in the kosmos.

It is not to be considered really that the scope of space is diminished but that the growing, expanding consciousness of the Thinker will so reduce the relative extension of it that illimitability will be swallowed up in its extensity. Consciousness, in becoming infinite in comprehension, annihilates the imaginary infinity of space. Accordingly, that which now appears to be beyond mental encompassment undergoes a corresponding diminution in every respect as the consciousness expands and becomes more comprehensive. *The mystery of space decreases as the scope of consciousness increases.* As the Thinker's consciousness expands the extensity of the manifested universe decreases. Thus the mystery of every aspect of kosmic life lessens, and fades away, as the intimacy of our knowledge concerning it becomes more and more complete. There is no mystery where knowledge is. Mysteriousness is a symbol of ignorance or unconsciousness, and that which we do not understand acts as a Flaming Sword keeping the way of the Temple of Reality lest ignorance break in and despoil the treasures thereof.

Figure 21 is a graph showing a sectional view of consciousness on all planes represented as seven concentric circles. This describes the analogous enveilment of the consciousness when it ensouls a physical body or when bound to the purely objective world of the senses. The overcoming of the barriers of reality, represented by the circumscribing circles is the work of the Thinker who is forever seeking to expand and to know. For only at its center, as symbolized here, is the consciousness at one with the highest aspect of kosmic consciousness and there alone is the mystery of space despoiled of its habiliments.

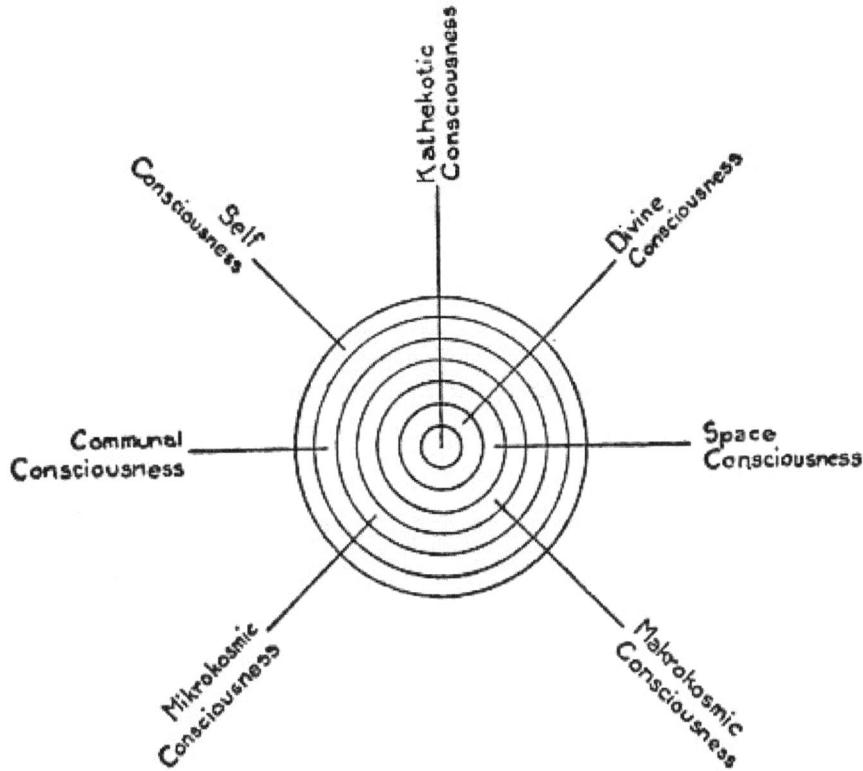

FIG. 21.—Septenary Enveilment of Consciousness

Accordingly, as consciousness or the Thinker is more and more divested of carnal barriers and illusions there develops a gradual recognition of the unitariness of spatial extent and magnitude; there arises the certain knowledge that space has but one dimension and that dimension is sheer *extension*. The Thinker's sphere of awareness is represented as if it begins as a point in space and develops into a line which divides into two lines, the boundaries of the space cones. Thus it may be perceived that the ancients had a similar conception in mind when they symbolized kosmogenesis with the dot (.), the line, and the circle with diameter inscribed, which together represent the universe in manifestation.

We realize the impossibility of adequately depicting the full significance of the inverse ratio existing between the extensity of space and the increscent inclusivity of consciousness by means of graphs; for neither words nor diagrams can portray the scope and meaning of the conception in its entirety. Yet they aid the intellect to grasp a ray of light, an intimation of what the Thinker sees and understands interiorly.

In this connection it is interesting to note the function of the ideal in the evolution and expansion of consciousness. The ideal has no perceptual value; it has no status in the world of the senses. It is unapproachable either in thought or action, and therefore, lies beyond the grasp of both the intellectuality and the vitality. It is indescribable, inconceptible and searchless; for the moment that we describe, define, or approach the ideal, either intellectually or vitally, in that moment it ceases to be ideal, but actual. It flees from even the slightest approach; it never remains the same; it cannot be attained, at least its attainment causes it to lose its idealty. It is then no longer the ideal. It is like an *ignis fatuus*; the closer we come to it the farther away it recedes. It hangs suspended before the mind like the luscious grapes which hung before the mouth of the hungry Tantalus. As the grapes and the water receded from his reach at every effort he made to seize them so the ideal remains eternally unseizable and unattainable. Whatever, therefore, is in our thought processes, or in our knowledge, that may be said to be *ideal*, does not really exist. The ideal is a phantom growing out of the nature and essence of the intellectuality. Its purpose is to lead merely the mind on; to allure it, to tantalize, and compel it to grow by exertion, by the struggle to attain, by the desire to overcome. In this respect, it serves well its function in the economy of intellectual evolution. It is a mysterious aspect of the original and eternal desire to live which is the kosmic *urge* present in all organized being and has its roots hidden in the divine purpose of creation.

Idealized constructions, then, are like Arabian feasts conjured up by a famishing mentality. They are like the dreams of a starving man in which he actualizes in phantom-stuff the choicest viands in abundant supply for his imaginary delectation. The mind that is satisfied never idealizes, never makes an idealized construction. It is only when an "aching void" is felt, when a longing for the realization of that which it has not arises within itself, when a feeling of distinct lack, a want, a hankering after something not in its reach, takes possession of the mind that it begins to idealize. That is why some minds are without ideals. It is because they are satisfied with what they have and can understand. They feel no hungering for better and grander things; they have no desire to understand the unknown and the mysterious; hence they do not idealize; they make no attempt to represent unto themselves a picture of that which is beyond them. Such minds are dormant, hibernant, asleep, unfeeling and unresponsive to the divine urge.

But the ideal is neither obtainable objectively nor subjectively, neither phenomenally nor really, so that when we come upon the ideal in our mode of thinking we have arrived not at a finality or end, but at that which is designed to lead us on to something higher, to nobler accomplishments and more extensive conquests. When we have devised idealized constructions, therefore, we should not therewith be content but should scrutinize them, examine and study them for their implications; for thereby we may discover the path and the guide-posts to a new domain, a new ideal, following which we shall, in time, come to a point in our search for the real where the fluxional is at a minimum; we shall reach that something which will admit of no further struggle—the last chasm between the phenomenal and the real —and standing on the bridge, consciousness, which engages the twain, shall have a complete view of that Sacred and Imperishable Land of Kosmic Realism where like a fleeting cloud of sheerest vapor shall be seen the phantom-ideal deliquescing and disappearing in the cold, thin air of the real and the eternal.

Since space is judged to be infinite by the intellect occluded in such clouds of illusion and hampered by such constrictive bonds of limitations, as it now endures, we have no right to conclude that the concept of infinity would still linger before the mind's eyes when the illusionary veil is removed; in fact, there is ample reason to believe, nay for the assertion, that the recession of the veil will reveal just the opposite of this illusion, namely that space is finite, and even *bounded* by the fringe of chaogenetic disorderliness. Either we perceive the real or we do not; either the pure *thingness* of all objects can be perceived or it cannot be perceived. If not, granting that there is such a thing as the real, it must be within the ultimate range of conceivability. It also seems reasonable that realism exists somewhere, and if so, must be sought in a direction inverse to that in which we find the phenomenal and the approach thereto must necessarily be gradual, continuous and direct and not by abrupt breaks, by twists and turns. The phenomenal must lie at the terminus of the real, and *vice versa*. So that by retracing the path blazed out by the real in coming to phenomenalization we shall perhaps find that which casts our shadowy world, just as by tracing a shadow in a direction inverse to that in which it extends we may find the object which projects it.

It is not out and beyond that we shall find the end of space; it is not by counting tens of thousands of light-years that the supposed limits of space shall be attained. The path of search must project in an opposite direction— not star-ward but Thinker-ward, toward the subtle habitation of the consciousness itself. We err greatly when we think that by measuring distances we shall encompass space; for that which we measure and determine is but the clouds caused by the vapor of reality. It is, therefore, not without, but within, in an inverse direction that the search must proceed. Going back over the life-stream, beginning where it strikes against the shores of solid objectivity, deeper and deeper still, past the innermost mile-stone of the self-consciousness, back into the very heart of the imperturbable interior of being where the Thinker's castle opens its doors to the Great Kosmic Self, from that open door-way we may step out into that great mystery of space—limited, yet not limited, multi-dimensioned, and yet having only one dimension, veritably real and fundamental, the Father-Mother of all phenomena. Here the great mystery of mysteries is revealed as the citadel of the universal and the ultimate real. In this citadel, the plane of kosmic consciousness, space loses its spaciousness and time its timeliness, diversity its multiplicity and oneness alone reigns supreme.

But the movement towards the center and circumference of space, after this manner, requires aid neither from the notion of space-curvature nor that of the space-manifold, except, indeed, only in so far as a state of consciousness or a degree of realism may be said to be a tridimensional manifold. The feeling that space is single-pointed, and yet ubiquitously centered, has been indulged by mathematicians and others in a more or less modified form; but they have imagined it in the terms of an indefinite proceeding outward until in some manner unaccountable alike to all we come back to the point of origin. It has been expressed by PICKERING when he says that if we go far enough east we shall arrive at the west; far enough north we shall come to the south; far enough into the zenith we shall come to the nadir. But this conception is based upon a notion of space which is the exclusive result of mathematical determinations and subject to all the restrictions of mathetic rigorousness. It requires that we shall allow space to be curved. This we decline to do for the reason that it is both unnecessary and contrary to the most fundamental affirmations of the *a priori* faculty of the Thinker's cognitive apparatus. It would seem to be necessary only that we should extend our consciousness backward, revert it into the direction whence life came to find that which we seek. By extension of consciousness is meant the ability to function consciously upon the various superkosmic spheres or planes just as we do on the physical. Yet it should be quite as easy to devise an idealized construction which would imagify the results of this ingressive movement of the consciousness as to represent the results of a progressive outward movement star-ward. Having done so the examination of them could be conducted along lines similar to those followed in the scrutiny of objective results.

What would it mean to the Thinker if he were able to identify his consciousness with the ether in all its varying degrees; what would it mean if he were able to identify his consciousness with life and with the pure mind-matter of the kosmos; and lastly, with the spiritual essence of the universe? What if his various vehicles of awareness were available for his purposes of cognition? What, indeed, if he could traverse consciously the entire gamut of realism and consciousness from man to the divine consciousness? Does it not appear reasonable that as he assumed each of these various vestures of consciousness, in succession, he would gradually and finally, come to a full understanding of reality itself? It seems so. This view is even more cogent when it is considered that the limitations, and consequent obscuration of consciousness are proportional to the number of vehicles or barriers through which the Thinker is required to act in contacting the phenomenal universe. Common sense suggests that freedom of motility is determined by the presence or absence (more particularly the latter) of bonds and barriers; that the less the number of such barriers the greater the scope of motility and consequently greater the knowledge.

PLATO evidently had this in mind when he imagined the life of men spent in cave-walled prisons in which their bodies were so fixed that they were compelled to sit in one prescribed position, and therefore, be unable to see anything except the shadows of persons or objects as they passed by. He conceived that men thus conditioned would, in time, suffer the diminution of their scope of consciousness to such an extent as to reduce it to identification with the shadows on the walls. Their consciousness would be mere shadow-consciousnesses the entire data of which would be shadowgraphs. So that for them the only reality would be the shadows which they constantly saw. A similar thing really happens to man's consciousness limited to the plane of the objective world. Things which are not objective do not appear as real to him, if they do appear at all. It is not that there are no other realities than those which appear to the egopsychic consciousness or that fall within its scope; but that this form of consciousness is incompetent to judge of the nature and appearance of those realities which do not answer to the limitations under which it exists. And so, with men whose data of consciousness or whose outlook upon the world of facts, or rather life, are confined to the narrow bounds of mathematic rigor and exclusiveness, there may appear to exist no realities which may not be defined in the terms of mathematics. Similarly, to the empiricist, used to measurements of magnitudes, weights, and rates of motion, there may also appear to be no realities which are not amenable to the mold of his empirical contrivances—the balance, the chromatic and the scalpel. All of these are shadow men constricted to the metes and bounds of shadows which they observe only because they are ignorant of the realities which lie without their plane.

Life has so many ways of exhibiting its remains to the intellect; and these remains have so many facets or viewpoints from which they may be studied, that nothing short of a panoramic view of all the modes of exhibition and of all the facets and angles of appearances will suffice to present a trustworthy and comprehensive view of the whole. Then, life itself is so illusive, so unseizable by the intellect that the testimony of all investigators are required to summarize its modes of appearance. And, therefore, eventual contentment shall be secured only when the mass of diverse testimonies is reduced to the lowest common divisor, and for this purpose the operations of every class of investigators must be viewed as the work of specialists upon separate phases, facets and angles of life's remains.

And so it is manifestly absurd for the empiricists, by taking note of the dimension, extent, quality and character of the shadows, or one single class of angles, to hope to predicate any trustworthy judgments about either the realities which cast the shadows or underlie the angles; because whatever notion or conception they may be able to gain must of necessity be merely fragmentary and entirely inadequate. Despite this fact, however, we still have the spectacle of men who, studying the sensible universum of space-content, endeavor either to make it appear as a finality in itself, or that the world of the real must necessarily be conformable to the precise standards which they arbitrarily set up in their examination of the objective world. It can be said with assurance that we shall never be able even so much as to approach a true understanding of the unseen, real world until we shall have changed our mental attitude towards it and ceased to expect that it shall necessarily be fashioned and ordered in exactly the same way as the world of our senses, or that it shall be understood by applying the same methods of procedure as those which we use in our examination of the phenomenal, sensuous world. It is a matter of logical necessity that, as there are no senses which can respond to the real, as there are no organs which vibrate in accord with the rates of vibration of the real, there can be no reasonable hope of understanding it by means of sensuous contrivances and standards.

Let the consciousness, therefore, be turned not outward, but *inward* where is situated the temple of divine life; let there be taken away the outward sheaths which enshrine the pure intelligence of the Thinker; let him grow and expand his sphere of awareness; let there be an exploration of the abysmal deeps of mind, of life and consciousness; for buried deeply in man's own inner nature is the answer to all queries which may vex his impuissant intellectuality.

CHAPTER IX

METAGEOMETRICAL NEAR-TRUTHS

Realism is Psychological and Vital—The Impermanence of Facts—On the Tendency of the Intellect to Fragmentate—The Intellect and Logic —The Passage of Space, the Kosmometer and Zoometer, Instruments for the Measurement of the Passage of Space and the Flow of Life— The Disposal of Life and the Power to Create—Space a Dynamic, Creative Process—Numbers and Kosmogony—The Kosmic Significance of the Circle and the Pi-Proportion—Mechanical Tendence of the Intellect and its Inaptitude for the Understanding of Life—The Criterion of Truth.

Kosmic truth has many facets. The rays of light which we see darting from its surface do not always come from the core. Often they are reflections of rays whose light stops short at the superfice; and these, in turn, are reflections of deeper realities. Thus the reflected light may be traced to its source by following the lead of external reflections. It is now known that moonlight, and perhaps, in many cases, starlight, are reflections of sunlight, if not of our sun, some other in the universe. But it is only at certain times and under certain conditions that we can see the sun which is the source of the other kinds of light. The stars which owe their light to suns are so many facets of sunlight. The moon is a facet of sunlight also. Facts are facets of truth. They are so many faces of eternal truth. They represent the many ways reality exhibits itself, or rather its effects, to the consciousness. When we, therefore, become aware of facts we have not in virtue thereof become aware of the reality which produces the facts. We have come to know only something of the termini of realism while the complexities and internal ramifications which lie between realism itself and these termini altogether elude our cognition.

Let us examine briefly an icosahedron, for instance. An icosahedron is a figure comprehended under twenty equal sides. These various sides are so many faces by means of which the figure presents itself to the consciousness. These faces, however, are not the real object. The figure may be examined by viewing it from any one of its sides; yet, by simply examining a single face, or any number of faces, less the total number, we arrive at no satisfactory knowledge of the magnitude or its substance. We must first become conscious of all the faces, holding them in mind as a composite picture, before we can even begin to have anything like a complete notion of the icosahedron. Then by continuing the examination we may find that the magnitude is composed of wood fiber or stone or metal, as the case may be. In this way we might carry the examination to indefinite limits and finally arrive at a very comprehensive knowledge of the icosahedron and yet be unaware altogether of the forces which have been at work in the production of the magnitude or of the reality which lies back of it.

Realism is psychological and vital. In essence it is mind, spirit, life. Yet these three are one. Mind is the outward vehicle of life; spirit is the form or the interior vehicle which life assumes in order to express itself. Realism, then, is life. Is the logician dealing with reality when he collects and coördinates the various modes of interpretation by which we learn to understand the symbolism of life? Obviously not. The data of logic are simply a collection of rules for interpreting concepts. It is a compendium of indices for the Book of Life. It is no more the book itself than a table of contents is a book. But logic occupies about the same category as does an index to a volume. A book, however, is more than its conventional contents. It is the thought that is symbolized therein. The book of life, accordingly, is the sum total of life's expressions; but it is not life itself. That is the subtle, evasive something which the contents of the book of life symbolize. Nature, both in her palpable and impalpable aspects, may be said to be the book of life wherein are recorded the movements, the expressions, and the diacritics of life. The whole is a magnitude of many facets (little faces). We shall have to know all the faces before we can say that we have a comprehensive knowledge of nature. For so long as we have only a fragmentary knowledge of the whole, so long even as we have merely a superficial knowledge of any aspect of nature, just so long will our knowledge be in vain. Just as it frequently happens that, on account of the partial view of things, we are led to make incorrect judgments concerning them, so when we come to make assertions about life or nature in general, we are apt to fall into the error of rendering judgments upon insufficient data. And it is not at all likely that judgments thus arrived at can possess true validity because it may happen, and undoubtedly does always so happen under the present limitations of human knowledge, that the very elements which are ignored or neglected in forming a judgment possess enough of virtue to alter the intrinsic value of determinations based upon otherwise insufficient data. Hence it develops not infrequently that our judgments repeatedly have to be changed in proportion as our data are made more and more comprehensive. Men searching eagerly for the truth sometimes allow themselves to be carried away by the enthusiasm of the moment which arises upon the discovery of a new facet of truth; but if all searchers were to bear in mind the fact that reality presents itself to consciousness in myriad ways and that there are innumerable facets all leading eventually back to the source of all they not so easily would be induced to jump to the conclusion that they had covered the entire ground. For when we have discovered a million facts, or many millions of them, about nature we may say that we have only merely begun and that what we have found is not to be compared with the totality even of the directly observable phases of nature.

Logic, therefore, deals with the symbolism existing between and among facets of truth, and not directly with truth itself, although the conclusions reached by the logicians may be true enough from an intrinsic standpoint. Logic is not truth, however; it is merely the consistence of relations and inter-relations between facts and among groups of facts. Truth is not established by logic; it stands in no need of the light of logic for its revelation; indeed, more apt than not is logic to obscure truth. Truth is its own proof; it is self-evident. Logic is a mere modeler of facts; it is static, immobile, fixed. All truly logical processes need a starting point, a foundation, a premise, a base. Truth, being eternal, mobile, dynamic, vital, needs no starting point; needs no foundation because it is itself fundamental; it requires no premise because no premise is comprehensive enough to encompass it. There is only one way of arriving at truth and that is not to arrive at all—just to recognize it without procedure. The fact that facts are, and the fact of their relations and inter-relations, their sequence and implications, can be arrived at only by logical processes. Life, in its passage through the universum of spatiality, carefully diacriticizes between the realm of facts and the domain of truth, marking each off from the other by unmistakable signs and barriers. Truth is perceived as an axiomatic, self-evident principle and no amount of logic could prove or establish its verity. Facts are intellectual creatures; truth is intuitional, vital. The intellect conceives the consistence of facts while the intuition recognizes truth—is truth, and therefore, follows in the wake of life as consciousness.

There is no permanence in facts and the intellectual recognition of their consistence. The discovery of a single new fact may destroy the consistence of a whole mass of previously correlated facts. Thus is revealed the miracle power of logic over facts. It can take a mass of facts, related or unrelated, mold them into hypotheses, endow them with a sort of interior consistency, and make these hypotheses take the posture of truth. Hence logic is often an effective mask which the intellect commonly imposes upon its material; but it does so instinctively and can no more escape the rigorous compulsion of this instinctive functioning than water can escape its liquidity. Wherefore, we conclude that true permanence abides alone in truth because truth is duration itself. For the foundations of the whole structure of facts in religion, science, art and philosophy which man has toilfully built up in the last million years might easily be destroyed or overturned by the discovery of some great fact or by appreciating the true value of truth. Let us suppose it should suddenly be realized by men that they are really and truly gods capable of creating and possessing all the other virtues, powers and capabilities which we are accustomed to impute to supreme divinity; and suppose that the fact of their omnipotence and divine omniscience always had been obvious but that men were so engaged upon details and the non-essentials of life and matter that they had not noticed nor realized it before, would not this realization make a vast difference in the character of our knowledge and the attitude which we would necessarily assume thereafter towards matter, life and the problems which they present? Would not it completely revolutionize our arts, our sciences and our philosophies? How much, then, of the facts of these would be left when the light of omniscience had been turned on—when truth itself could be perceived and interiorly realized? Not much, to be sure. We should undoubtedly have to dispense with the entirety of our fact-mass, for it should then be entirely useless and meaningless in the light of the resplendent omniscience of truth. As at present constituted consciousness is focused upon the material plane for the purposes of superficial observances. But if the focus of consciousness should be changed so as to reveal conditions upon what must be a higher and more interior level, the aspect of things would be entirely changed and the whole of our theory of knowledge would have to be reconstituted. It is conceivable, yea obvious, that the stern reality of being is far removed into the Great Interior of that which is; and there is a point in the path to the interiority of being where there is no illusion, no appearance, indeed, nothing but the cold, illuminating body of reality itself. It must appear also that along the journey interior-ward there are many apparent levels or planes, each of which requires a new focus. It is unreasonable, then, to suppose that the conclusions arrived at as a result of purely logical processes, confined to the lowest levels of reality, are pertinent and valid for the entirety of realism which is neither of mathematical nor logical import. For instance, if we take the purely axiomatic assertion: x equals x, the intellect is at once certain that this is so, and cannot be otherwise, and yet a proposition of this kind is purely conceptual, conventional and arbitrary. x may also equal 1, 2, 3, 4, or any other quantity. Then, if each x in the above equation be replaced, by say, a horse, there immediately arises a difficulty. For it is not possible to find two horses which are in all respects mutually equal. So that as soon as we pass from the conceptual into the actual, whether on the side of objective reality or that of absolute reality, the validity of the axiom is immediately exposed to serious questioning. The truth of the matter is that on both sides of the conceptual it is always found that there is a variance from the standards set up by the conceptual, this variance being more marked on the side nearest to absolute reality than on the side of objectivism. Objectively, the conformity of the sensible with the conceptual is of such approximation as to lend trustworthy utility to the conceptual in its application to the sensuous. Thus by simply eliminating the vital factors from our equations we are enabled to proceed in a reasonably safe manner with our judgments. Really, however, no such approximate congruence can be found; for on the side of reality we are dealing with an indivisible something—something that is eternally and absolutely unitary in its constitution while when we transfer the scene of our observations to the objective world we discover a contrary situation. Here we are everywhere beset by diversities, multiplicities and dissimilarities. This is so because the intellect naturally tends toward the objective where it finds a most comfortable atmosphere for its operations. The conceptual is related to the objective as a train of cars is related to the railway. That is to say, the constitution of the intellect is such that it finds its most facile expression in the objective world and is about as comfortable in the domain of realism as the same train of cars would be on the ocean.

The intellectuality is designed to deal with facets of truth; it is made to manipulate segments, parts, fractions, and cannot chart its way through a continuum such as reality. Being constitutionally a conventionality of the Thinker's own contrivance, and arising out of the subtle adaptation of his vehicles to the environment afforded by the sensuous world, it can only find congruence in that conventionality which is the instrumentality of a higher intellectuality expressed in a diversity of forms, into which reality divides itself for manifestation. The human intellect is, therefore, the bridge over which is made the passage from the individual consciousness to the All-Consciousness; simultaneously, the medium whereby the physics of the brain are converted into the psychics of unconsciousness. It may be likened to a pair of specially constructed tongs which are so formed as to fit exactly the objects which a higher intellectuality has made. It is without the province of the intellect to take note of what intervenes between physics and psychics; it is always oblivious of interstices while taking cognizance of objects or things. In this respect, the intellect is much like a steerage passenger on board an ocean liner who sees only his port of departure and port of arrival, knowing nothing in the meantime of what happens during the voyage, nothing of what the other passengers on the upper decks may experience and taking no part in any of the passing show until he lands. So that the passage of the intellect from fact to fact is an altogether uninteresting voyage; it may as well be made unconsciously, and to all intents and purposes, is so made.

Accordingly, the advocates of n-dimensionality find it quite impossible to predicate anything whatsoever of the passage, say, from tridimensionality to quartodimensionality. They find themselves at ease in tridimensionality and have even contrived to find pleasant environs in the four-space having made therein such idealized constructions as will afford ample hospitality to the intellect. But the questions as to how the passage from the three-space to the four-space is to be made and how the intellect shall demean itself during the passage have been completely ignored and, therefore, left unanswered. What, then, shall be said of an explorer who says he has found a new land and yet can give no intimation as to how one may proceed to arrive at the new land, what changes are to be made en route, nor the slightest suggestion as to the direction one should take in setting out for it? It is not likely that the report of such an explorer, in practical life, would be taken seriously; and yet, there are those who, relying utterly upon similar reports made by certain enthusiastic analysts, dare to place credence in their asseverations. Not only have they given wide credence to these reports, but have, indeed, sought to rehabilitate their own territory in accordance with the strange descriptions given by unhappy analytical explorers. Now the question of greatest concern, granting for the nonce that there is such a domain as hyperspace, is the *passage*. How shall we make the passage? Or, is the passage possible? In vain do we interrogate the analyst; for he does not know, nor does he confess to know. Evidently it is impossible for him to know by means of the intellect alone; for the intellect not being fitted to take cognizance of the "passage," but only the starts and stops, has no aptitude for such questions. Hence, what seems to be the most important phase of the entire question will have to remain utterly inscrutable until the intellect nourished by the intuition shall be aroused from its lethargy and brought to a certain high point of illumination where it, too, may take note of the passage.

Space is the path which life makes in its downward sweep through all the stages of pyknosis or kosmic condensation by virtue of which it accomplishes the engenderment of materiality as also the path marked out by it in its upward swing whereby it accomplishes the spiritualization of matter. It is the kosmic order which life establishes by means of its outgoings and incomings. When we look out into space we perceive that which is a dynamic appearance of life itself, and not a pure form. Nothing that is a pure form can exist in nature and in as much as space is not only indissoluble from nature but partakes of its very essence it cannot be said to be a pure form. The intellect, however, prone to follow the grooves laid out by pure logic, never fails to seek to make everything that it contacts conform to these logical necessities. But, if the analyst were to make careful discrimination as to the respective categories—that into which life falls and that in which the intellect is forced by its nature to proceed—he not so easily would be led into the fault of attempting to shape realities upon models which being strictly conventional were not meant for such uses. But neither the logician nor the mathematician can be condemned for such generosity if such condemnation were justifiable. For they everywhere and at all times insist upon *realizing* abstractions and *abstractionizing* realities, and they do this with an *insouciance* that is at times surprising. Yet it is in this very vagary that is discovered the true nature of the intellect. There is a sort of dual tendence observed in the method of the intellect's operation. A polarity is maintained throughout: the abstractive and the concretional. It vacillates continually between the abstract and the concrete and no sooner has it found a concrete than it begins to set up an abstract for it; and *vice versa*—as soon as it is has constructed an abstract it immediately seeks either its concrete or sets out to hew some other concrete into such shape as will fit it. And between these two extremes numerous excuses are found for exercising this peculiar characteristic, and that too, without regard to consequences. It would seem that the intellect, in thus functioning, was really engaged at a sort of sensuous play out of which it derived an intense and not altogether unselfish pleasure.

Of course, it must be granted that diversity has its specific and withal necessary uses in that it affords a field for the operation of human intellectuality and represents the adaptation of the kosmic intellect to the human for the purposes of evolution. This adaptation while necessary for the intellectual development is, however, not an end in itself. It is merely a means to a higher purpose. In fact, if we regard materiality as a deposit of life, carried by it as a kind of impedimentum, and consciousness, which *is* life, as being identical with the intellectuality which makes these adaptations, there should be no grounds for the statement that the one is adaptable to the other at all. And as this is really the view which we assume it would perhaps be more strict to regard the adaptation as subsisting between the human intellect and materiality both of which having been constructed by kosmic intellectuality. Pursuant to the diversity of uses to which materiality lends itself there arises in the intellect a supreme tendency to segment, to break up into separate parts, to multiply and diversify. It is not content unless it is at this favorite and natural pastime. It delights in taking a whole and dividing it into innumerable parts. This it will do again and again; because all its muscles, sinews and nerves are molded in that mold and can no more cease in their tendency to fragmentation than can the muscles of a dancing mouse cease in their circular twirling of the mouse's body. Yet, in this it is but creating a well-nigh endless task for itself—which task must be performed to the uttermost. But in its performance, that is, in the intellect's complete understanding of the diversity of parts, in the knowledge of their relations and inter-relations and in their synthesis, it may arrive at that one ineluctable something which is called *unity*. And so doing, become ultimately free.

In view of the foregoing, it is not surprising that the intellect should have, finally, fallen upon the notion of *n*-dimensionality. It has come to that as naturally as it has performed its most common task. Left alone and unhampered in its movements, it has simply followed the lead of the Great Highway through the domain of materiality. And now it has arrived at a stage where it thinks it has succeeded in fractionalizing space. Time has long ago yielded to fragmentation, been divided into minute parts and each part carefully measured. Space, not having a visible indicator like time to denote its passage or parts, suffered a long and tedious delay before it could boast of a measurer. As the sun-dial measured time in the past and became the forerunner of the modern clock so *n*-dimensionality measures space for the mathematician. What more practical instrument for this purpose may yet be devised is not ours to prophesy; yet it is not to be despaired of that some one shall find a suitable means for this purpose. Seriously, however, it is not without possibility that should some subtle mind devise an instrument for marking the passage of space as we have for denoting the passage of time a great stride forward would be accomplished in the evolution of the human intellect. For the general outcome of the intellect's attention being turned to the *passage of space* would undoubtedly be to recognize not only its dynamism but its *becoming-ness*, as a process of kosmogenesis. Because such an instrument would have to be so constructed as to take note of the movement of life, and for this reason, it would have to be extremely sensitive necessarily and keyed to the subtleties of vitality and not to materiality. Mathematics shall have failed utterly in the utilitarian aspects of this phase of its latest diversion if it do not justify its claims by crowning its work in the field of hyperspace with a "Kosmometer," an instrument devised for the measurement of the movement of space or a "Zoometer," an instrument devised for the measurement of the passage of life. We should like to encourage inventive minds to turn their attention life-ward and space-ward with the end in view of constructing such an instrument. When once we have learned accurately to measure life we shall then be able to dispose of it—to *create*. It is not doubted that if ever humanity is to arrive at that point in its evolution where it can understand life; if ever it is to attain unto the supreme mastery both of vitality and materiality and to come to the ultimate attainment of divine consciousness (all of which we confidently believe to be in store for humanity) it must be accomplished after this manner: first, by syncretizing materiality with vitality, and then, by intuitionally recognizing the truth of the implications of the syncretism.

The history of consciousness in the human family is identical with the history of man's conquest over matter and physical forces. And this is clearly indicated in the incidentals contingent upon the toilsome rise of the *genus homo* from the earliest caveman whose status denoted a comparatively negligible transcendence of material forces, to the present-day man who has gained a markedly notable conquest over these forces. Always consciousness seeks the means of adequately expressing itself in the sensible world. And to this end it engenders faculties, organs and processes in the bodily mechanism, and, in matter, devises instruments of application whereupon and wherewith it may test, analyze, combine and recombine the forces and materials it finds. The unlimited range of expressions lying open to the consciousness makes it necessary continually to devise higher and higher grades of appliances to meet its needs as it expands. It will not be gainsaid that the telescope has served actually to lay bare to the consciousness an immeasurable realm of knowledge nor that the microscope, turning its attention in an opposite direction, marvelously has enlarged and enriched our knowledge of the world about us. And similar declaration may be made anent almost every invention, discovery and conquest which man has made over natural phenomena. Thus, by externally applying mechanical implements to the subject of his consciousness, man has extended actually his consciousness, his sphere of knowledge; has greatly enhanced its quality, and, in the process, has urged the intellect to endeavors that have wrought its present unequaled mastery of things. Nor have the spiritual aspects of our advance along these lines been the least notable. For these have enjoyed the essence of all that has been gained in the process and have, therefore, kept pace with the onward movement of the intellectual consciousness. But heretofore no advance has been made as a result of methodic or reflexive determinations. That is, men did not set out from the beginning, equipped with foreknowledge of what their efforts would bring, to develop the present quality of human consciousness. They simply worked on, their attention being absorbed by the problems that lay nearest and demanded earliest consideration. So the advance has come as a resultant of man's close application to his ever-present needs—shelter, clothing, food, protection and other preservative measures—and it has come naturally and inevitably and without prepense. Nevertheless, if man, knowing what to expect from the syncretization of matter and mind, after this fashion, should set out deliberately to accelerate the intensification, expansion and growth of his consciousness, there is no doubt but that the consequence would be most far-reaching and satisfactory.

But the path that leads to this grand consummation does not lie in the direction of diversity; it lies in the opposite direction. In vain, then, does the intellect fractionalize in the hope that by doing so it shall come to the solid substructure of life; in vain does the analyst segment space into any number of parts or orders; in vain does he ask how many and how much; for by answering none of these queries will he find the satisfaction which he vaguely seeks.

If it be true that it is not by analysis but by synthesis that the true norm of life, and therefore, of reality shall be found it is futile to entertain serious hope of finding it in any other way. As a perisophism or near-truth, then, *n*-dimensionality takes foremost rank. And this is so for the reason that when we proceed in the direction of multiple dimensions, that is, one dimension piled upon another dimension or inserted between two others we are traveling in a direction which, the more we multiply our dimensions, leads us farther and farther away from the truth. This is a simple truism. If we take, for instance, a wooden ball and cut it up into four quarters, and divide each one of these quarters into eighths, into sixteenths, thirty-seconds, sixty-fourths, etc., indefinitely, we shall have a multiplicity of parts, each one unlike the original ball. But from no examination of the multipartite segments can we derive anything like an adequate conception of the original ball. Something, of course, can be learned, but not enough to enable the rendering of a correct judgment as to the nature, size, shape and general appearance of the ball. But this is precisely what happens when the analyst divides space into many dimensions. He cuts it up into *n*-dimensional parts and the more minutely he divides it into parts the more remote will each part be in its similarity to the original shape and form of space, and the farther away from the true conception of the nature of space he is led thereby.

Now, *n*-dimensionality or that phase of metageometry which regards space as being divisible into any number of dimensions or systems of coördinates is a direct and inevitable product of that tendency of the intellect to individuate and to singularize phenomena. Biologically speaking, it is a peculiarity which harks back to the time when life was manifested through the cell-colony and when the individual cells began, because of increasing consciousness, to detach themselves from the colony and set out for themselves, and thus each intellect recapitulates in its *modus vivendi* the salient tendencies of phylogenesis. Let it suffice, then, to point out that this universal tendency to segment and fragmentate which rigorously characterizes intellectual operations upon every phenomenon with which it deals is a culmination of the primordial tendency among cells to divide, inasmuch as this phase of cell life must be the work of the kosmic intellect. The natural inference is that from the extreme of individualization there shall be a gradual turning, whether of the intellect *per se* or of the intellect joined to the intuition does not matter, towards that other extreme of *communalization*. And from this latter shall grow up, as one of the inevitable and ineluctable tendencies of the Thinker's consciousness a torrentious movement in human society towards coöperation, brotherhood, mutuality and union in everything. So that whereas in the past and at the present time the intellect has been developing under the dominant note of individuality it will then be coming gradually under another dominant note —*communality*. The result of this development will be the unification of all things, and instead of many dimensions of space, many measures of time, and a general diversification of all phenomena, we shall come to the only true notion of these things and realize pragmatically the true value and extent of unity in the universe.

It is admitted that the intellect, in treating objects singly and dealing only with the starts and stops of a movement, is withal loyal to the kosmic order, design and purpose which have priorly characterized manifested phenomena by segmentation. And in this loyalty it has been following merely a natural lead which, while admitting of the widest development and experience, nevertheless at the same time beneficently obscures the underlying reality in order that in its adaptation to the sensuous world the intellect might have the greatest freedom for the development suited to the given stage of its evolution. But in thus admitting the natural congruence between the intellectuality and the phenomenal or sensuous we do not thereby unite with those who already believe that this kosmic agreement is the *ne plus ultra* of psychogenesis. On the other hand, it is maintained that this is merely a phase of psychogenesis which shall be outgrown in just the same measure as other phases have been outgrown. And notwithstanding the fact that judgments of the intellect with respect to inter-factual relations or the ens of facts themselves are as valid as its judicial determination of self-consciousness, no more and no less, we are, by the very rigor and exclusiveness of this logical necessity and inherent limitation, led to view the intellect's interpretation of phenomena as partial and fragmentary; for the reason that the necessitous confinement of its understanding and interpretative powers to fact-relations quite effectively inhibits the use of these powers for the contemplation of the deeper causative agencies which have operated to produce the phenomena. But it is apparent that just as the transmuted results of other phases of psychogenesis are now being utilized as a basis for the efficient operation of the intellect in the sensuous world, thereby enabling the attainment of a very high mastery over matter, so will the functional dynamism acquired by it in the pursuit and comprehension of diversity serve well when, in later days, it has acquired the power to deal directly with reality, to *create* and dispose of life just as the kosmic intellect has and is now using it in the execution of the infinite process of *becoming* through which creation is proceeding. It would seem that the necessary prerequisite to the development of any higher functional capability is that the intellect should be capable of disposing of innumerable details, indeed the totality of kosmic detail, before it can come wholly into the power and capacity to understand and manipulate life. Furthermore, it appears that the acquirement of this power quite necessarily has been delayed awaiting that time when, dominated by the intuition, the intellect shall have attained the requisite managerial ability for marshaling an exceedingly large number of details.

The supreme tendency of life is expression. And this expression, singularly enough, reaches its most perfect phenomenalization by means of that movement which results in the multiplication of forms. Despite the fact, therefore, that the comprehension of reality involves a gradual turning away from the exclusive occupation of organizing a multitude of separate and apparently unrelated facts to a monistic view which at once recognizes the unitariness and co-originality of all things, of life, mind and form, the intellect will need the training and development which come from the mastery of diversity. It is, then, not difficult to perceive the wise utilitarianism of the present schematism of things as shown in the universal tendency in the intellect to devote itself exclusively to parts or segments of truth.

Whenever an individual intellectuality, on account of prolonged thought and the consequent inurement of the mind to higher and higher vibrations of the kosmic intellect, brings itself to such a high point of sensitiveness that it can receive so much as an intimation of some great truth, it begins to sense, in a more or less vague way, something of the substance and general tendence of the underlying reality of that which foreshadows its appearance. Then, confounded by the multiformal characteristics of kosmic truth because of the fact that it presents itself in such numerous ways and forms, men often are induced to attempt the reformation of all facts, or a great mass of kindred facts, in accordance with the newly-found fact or principle. They forget evidently that no fact in the universe can be at variance with any other fact and still be a fact. So that in the totality of facts every separate and distinct fact must be congruent with every other fact forming a beautiful, harmonious and symmetrical whole; but often the whole is made to suffer in the attempt at making it conform to the substance of a mere intimation. Moreover, it is conceivable that even the totality of facts may lack a rigid conformity with reality in all its parts and that having compassed the entire mass of facts one may fall short of the understanding of realism.

This is practically what has happened in the mind of the metageometrician who having received an intimation as to the real nature of space as that whose center is everywhere and yet nowhere and whose nature is psychological and vital rather than mathematical and logical, misses the great outstanding facts and clings to the intimations which he experiences as to the nature of space. He, therefore, concludes that the form of space is that of a flexure or curve. There is a valid element in the notion of the curvature of space but not enough of truth wholly to validate the notion. Since the very reality of space is a matter which can be determined only by the conformance of the consciousness with it in such a manner as to render the conception of it entirely unintelligible to the intellect except in so far as it may be able to identify itself with the space-process, there is much room for the serious questioning of the mathematic conclusion upon the grounds of its fragmentariness if not entirely upon the basis of its invalidity. Wherefore it may be seen that any search for either the center or the extreme outer limits which proceeds in a manner conformable to the external indications of the intellectual order is vain, indeed. Although it is undoubtedly true that the attainment of a central or frontier position in space does not involve any lineal progression whatsoever, the same being attainable, not by progression nor by overcoming distances, but by a subtle adjustment, yea, a sort of attachment of the consciousness to the order of becoming which binds the appearance of space, wherever one may be, it is nevertheless difficult and painful for the intellect to grasp the totality of this truth at one sweep. Indeed, it is not possible for it, alone and unaided by the intuition, to grasp it at all. Hence, the mathematician who depends entirely upon the deliveries of the intellect which conform, in their passage from the conceptual to the written or spoken word, to all the rigors of mathetic requirements, fails utterly in perceiving the magnitude of this conception and all its connotations; he fails because his prejudices and the woof and warp of his intellectual habits prevent his assuming a sympathetic attitude toward it and thereby precluding at the start any calm consideration of it. And not only is this true of the mathematician but of all those whose endeavors are confined to the plane of purely sensuous and logical data. It would, therefore, appear that our entire attitude towards things spatial must be changed before we can even begin to perceive the reality which is really the object of all researches in this domain. But, on the surface, there is after all little difference between the ultimate facts involved in these two totally different conceptions. Mathematically speaking, all progression eastward would terminate at the west, and *vice versa*; and the same would be true regardless of the point from which progression might originate. Always the terminus would be the opposite of the starting point. Then, too, it might be said that if we sought the space-center we should arrive at the circumference. The difficulty with this view is that there is a very remote, though important, connection between it and the truth of the matter. But the partiality of this view, and the absence of either experience or intuition to intimate a more reasonable view, serve effectively to buttress it as a hypothesis acceptable to many. Thus it is ever more difficult to supplant a near-truth than it is to gain credence for the whole truth. On the other hand, according to the view which we maintain here, it is quite true that the seeking of the kosmic space-center will reveal the circumference; that the search for the nadir will uncover the zenith; the east effloresces as the west, and a northward journey will wind up at the south, etc., but in quite a different manner from that which the mathematician has in mind when he postulates the curvature of space. Our view involves no space curvature nor any other spatial distortion. *It deals with space as reality, as a dynamic process, a flux which, like the sea, is continually casting itself upon the shores of chaos and falling back upon itself only to be recast against the rock-bound coast of its chaotic limits.* Now, that which falls back upon itself and rolls in a recurrent movement upon its own surface is *life* which, in its recession is the natural and kosmic limitations of itself, generates matter in all its varied expressions. Space, in its extensity, cannot transcend life; for it is the path which life makes in its *out-coming*, its manifestation. Of the chaotic fringe which circumscribes the manifested universe it is absurd to say that it is vital or psychological in any sense of these terms. For notwithstanding the fact that out of its very substance are engendered life, intellectuality, spatiality and materiality, it is nevertheless none of these in its primary essence. It is Chaos-Kosmos; because from its content the kosmos is evolved, and it still remains; it is chaos-spatiality; chaos-materiality; chaos-intellectuality; chaos-geometricity; because these are engendered by the movement of life in chaos while at the same time there remains a residuum of the chaogenetic substance which constitutes the limitations of all these subsequent processes. In this sense, the chaogenetic fringe becomes the limits of the manifested universe so that it would appear that all those major processes outlined above are finite manifestations of the eternal chaos. But none of those possibilities of motion which are found in these major movements of the kosmos can be logically said to exist in chaos. It is the embodiment of everything that is the opposite of those qualities which may be found in them, that is, in materiality, vitality, spatiality, intellectuality and geometricity.

Apropos to this phase of the discussion let us examine briefly one of its most significant implications, both mathematical and kosmic, which arises out of the fact that space is an engendered product of life that is bound by the fringe of chaos which sustains and limits it. The chaotic fringe plus manifesting kosmos constitute the absolute magnitude of the kosmos. The manifestation factor is complemented by the chaos factor and together the two define the *full* universe. Kosmogony is the universal movement of all kosmic elements or factors in diminishing the chaotic complement and reducing it to kosmic order or geometrism. It is undoubtedly impossible to determine mathematically the exact volume of either complement or the ratio of the one to the other; yet it is conceivable that the chaotic fringe is greater in extent than the ordered portion of the kosmic uni-circle or universe. It is even conceivable that the difference, upon the basis of the meaning of the Pythagorean Tetragrammaton and the view outlined in the Chapter on the "Mystery of Space," is as seven to three wherefrom the conclusion might be drawn that the universe has yet seven complete stages more or less of evolution before the close of the Great Cycle of Manifestation when the fringe of chaos shall have been totally used up in the work of creation. But for those who may experience impatience at the infinitude of the process when viewed in this light the terms may be reversed and the difference may be conceived as the ratio of three to seven wherefrom the conclusion would follow that the kosmogonic process is seven-tenths complete, as it will not vary the seeming infinitude either way it may be determined. The notion, despite its speculative character, offers an explanation of otherwise inexplicable conditions, and, on account of its profound connotations, may even be found to be productive of the highest good in its equilibrating influence upon our mode of thinking.

In any event, there does appear to be a subtle relation subsisting between true numbers and kosmogony. Number is a phase of the kosmogonic movement, a measurer of the intellect and the establisher of the geometrism of space, answering tentatively to the numericity of pure being. In fact, being actually expresses number and number itself is an evolution and not a thing posited once for all as a pure, invariable form in the universe. It is, like the kosmos, in a state of becoming and there may yet appear to our cognitive powers a whole series of new numbers pure in itself and altogether conformable to the conditions reigning at the time.

The symbology of the circle, in all times recognized as the true symbol of the kosmos in eternity, of eternity itself, of the archetypal, of space, duration and Ultimate Perfection, is replete with profound significations. But it should be understood that the circle is a symbol of the *perfected universe* and not the universe in a state of evolution. It symbolizes perfection, completion and the ultimate union of the manifesting with the archetypal which results in the crowning deed of Perfection. The circle is, therefore, not a symbol of the universe as it now stands; it does not represent a snapshot view of the kosmos but the universe as a *full*. It cannot be a *full* until it has attained the *ne plus ultra* of completion; for a kosmic full is that state to be attained by the manifested kosmos upon the termination of all the fundamental processes now in operation. But it is this state that the circle really represents, and by virtue of which it possesses its intrinsic qualities and also in virtue of which the intellect recognizes these qualities. The properties of understanding and recognizance in the intellect are veritably fixed by the *status quo* of the universe during every stage. That is, the focus of the intellect, like the focus of a chromatic lens, is adjusted by the fiat of the nature and eternal fitness of things to correspond exactly with every state through which the kosmos itself passes. This is one of the obvious implications of the phanerobiogenic behavior of the kosmos and is necessarily resident in the notion of the genesis of space and intellectuality as consubstantial and coördinate factors.

Wherefore the more cogent is the reason for the belief that the inherent qualities of the kosmogonic fundamentals; as, vitality, materiality, spatiality, intellectuality and geometricity, are true variants, and that their variability is proportional to the progress of these major movements toward the ultimate satisfaction of the original creative impulse. May it not be, therefore, that the indeterminate character of the ratio of the diameter to the circumference (3.1415926 ...), is due to causes far more profound than the crudity of our micrometers or the mere supposed fact of the circle's peculiarity? May it not also be true that the *pi proportion* shall become a whole number, and in its integration, keep apace with the perfecting process of the kosmos, diminishing, by retrogression to one or increasing, by progression, to ten which, after all, is essentially unity, being the perfect numeral? It is not without the utmost assurance that these queries will be categorically questioned by the orthodox, creed-loyal, strictly intellectual type that we sketch these implications, but it is felt to be an urgent duty to remind all such that the most effective barrier to realization in the field of philosophy is an intolerant attitude towards all lines of thought which suggest the impermanence of conditions as we find them in the kosmos at the present time. The fact is that our lives are so distressingly short that we have neither time nor opportunity to watch the changing moods of the kosmos nor discern the gradual reduction of mere appearance to the firm basis of reality, and accordingly, the intellect tenaciously clings to those notions which it derives from the instant-exposure which the lens of intellectual conceivability allots to it. Once the view is taken it is immediately invested with everlastingness. This everlastingness is then imputed to the kosmos in that particular pose, attitude or state. Always the intellect beholds in that passing view, snatched from the fleeting panorama of eternal duration, a picture of itself which it mistakes for the reality of the not-self.

The inclination of the axis of the earth toward the plane of its orbit is approximately twenty-three and one-half degrees. No well-informed astronomer, however, doubts now the fact that this ecliptic angle is being gradually lessened; because, as a result of centuries of observation, it has been found to be decreasing at the rate of about 46.3 seconds per century. Yet no intellect is able to perceive in any given lifetime the actual decrement of this angle. It is only by careful measurements after centuries of waiting that a difference can be discovered at all. Thus it may even be so with the ratio of the diameter to the circumference of a circle, the only difference being that it has not yet been determined whether there is a *decrement* or an *increase* in the size of the ratio.

The *pi* proportion is, then, a register or measurer of the slow, measured approach of the manifesting kosmos to the standard of ultimate perfection. Therefore, and in view of these considerations, we may not hesitate to confirm our belief in the validity of the notion that it actually and literally expresses the key to the evolutionary status of kosmogony. The mathematical determination which limits it as an unchangeable, inelastic quantity is, consequently, only partially true and leads to the inclusion of this quantity under the category of mathematical near-truths, for such it appears to be in spite of its rigorous establishment.

The formal topography into which the intellect spreads when seeking the ideal and the abstract is not a condition which is derivable from the real essence of life or matter, but, on the other hand, is a product of the intellect itself partaking of its nature rather than of the nature of reality. There is, therefore, a very important distinction to be made between all deliveries of the intellect and the realism both of the objects and conditions to which the intellectual deliveries pertain. One of the most marked peculiarities of the human intellect is the fact that it always unavoidably stamps its own nature and features upon every datum which passes through it to the consciousness. The utmost importance attaches to this phenomenon, for the reason that it points to the necessity of carefully scrutinizing intellectual deliveries and the making of allowances for those ever-present characteristics which the intellect superimposes upon its data. Perhaps the inherent colorific quality which it imposes upon our knowledge would be better understood if a similitude were indulged at this juncture. The intellect may be likened to a color-bearing instrument which, when it has once handled an object, leaves forever its own color transfused into every cell and fiber of the object so that when the same object is presented to the consciousness for purposes of cognition it bears always the same peculiar marks and colorations which the intellect, in its manipulation of it, places thereon. In this respect the intellect may also be said to be like a potter who has but one mold and that of a peculiar formation. Hence, whatever wares it presents to the consciousness will invariably be found to be molded in conformity with that particular mold. If it were possible to view reality or the essential nature of things the difficulty which now the intellect lays in the path of direct and uncolored cognition would be obviated; for then there no longer would be any necessity of viewing things as they are colored or molded by the intellect. The intuition, being a process of pure consciousness, will, when it has arisen to a position where it may dominate the intellect as the intellect now dominates it, so modify this tendency which we see so ineradicably bound up in the very nature of the intellect that the apparently insurmountable difficulties which it has interposed between mere perception and a direct cognitive operation will be quite completely overcome. Thus, in the above, is discovered another obstacle which posits itself between the notion of space as reality and the intellectual determination of it which the mathematician examines and to which his consciousness is necessarily limited. Furthermore, it may be perceived also how easily the mind may be deluded into thinking that the intellectual notion which it entertains of space is necessarily correct, when obversely, it is simply examining a concept which has been remade by the intellect into a form which is not at all unlike its own peculiar nature, and therefore, as much short of reality as the intellect itself is. Similarly, if the mathematical mind succeed in catching a glimpse of the reality of space in the form of an intimation, which, in itself though fragmentary, is nevertheless true, its consciousness is finally deprived of the true validity thereof simply because of the behavior of the intellect in its manipulation of it. The importance of these intellectual difficulties cannot be over-estimated for they furnish the grounds for the ineptitude of intellectual determinations made in a sphere of motility to which the intellect is a stranger. And this fact will appear more evident when it is perceived that quite the entire content of human knowledge has been thoroughly vitiated by them. So that only in those very rare moments which (in a highly sensitive mentality) enable the intuition to gain a momentary ascendancy over the intellect is it possible for the Thinker to catch hold of realism itself, and project the truth of what he sees into the lower, intellectual consciousness. But so small is that portion of our knowledge which owes its origin to the intuition that when compared with the totality of that which we seem to understand it is well-nigh negligible. And then, when it is considered that at present there is no way of conceptualizing adequately the intuitograph so as to make it propagable the insignificance of this form of knowledge is even more notable. It can now be seen in how large a measure the notion of the curvature of space is merely an intellectual translation of a true intuition into the terms of the intellect which, in the very nature of the case, can only approximate the truth because of its colorific habits.

A similar declaration may be made of that other datum of metageometrical knowledge which postulates the ultimate convergence of parallel lines. In fact, what has been said as to the perisophical nature of the notion of space-curvature will apply with equal force to the idea of parallel convergence since the latter is a derivative of the former. But there is yet another consideration, apart from the colorific influence of the intellect, which, although it partakes of the nature of this quality, is nevertheless a near-truth of quite a different order. This may be better understood by referring to the *graph* showing the inverse ratio of objective space to the consciousness.[28]

Let us suppose that the *graph* may also represent the Thinker's outlook into the world of spatiality. It then appears that, because of that movement of consciousness in its pursuit of life which, as it expands, makes the objective world to appear to be diminished in proportion to the extent of its expansion, it is quite natural, under such circumstances, that parallel lines drawn anywhere in the limits of the objective world should seem to come to a point in the ultimate extension of themselves. While this *graph* is not meant to depict such a view, it may be found nevertheless, to be a true delineation of the topography of that state of mind into which the metageometrician brings himself when he visualizes space as *curved*; for there is no doubt but that a state of intellectual ecstasy, such as that in which the mind of the metageometrician must be functioning in order to perceive space in that form, is quite different from the normal and, therefore, in need of a different topographical survey. But, if we grant that in the creational aspects of space there is conceivable an ever-present tendency to convolution, or a rolling back upon itself, it is imaginable that parallel lines inscribed either upon its surface or in its texture need not necessarily meet but maintain their parallelism regardless of the complexity of the convolutions. The convergence of parallel lines is much like a tangent in the outgrowth of the idea from the notion of space-curvature. The more a tangential line is extended the farther away from the circumference it becomes and consequently less in agreement therewith. The more subsidiary propositions or corollaries are multiplied the more remote from the truth the determinations become and especially is this true of the hypothesis of space curvature.

In the notion of the manifoldness of space, by virtue of which it is conceived as existing in a series of superimposable and generable manifolds of varying degrees of complexity, are discernible traces of that intuitional intimation which underlies the assertion that because of the necessary phenomenalization of reality for the purpose of manifestation to the intellect it appears to exist in a series of separate degrees, each one more refined and subtle than the preceding one and requiring a more highly developed species of consciousness for its comprehension. In other words, that intuitional glimpse of the essential character of reality, as viewed by the human consciousness, which impinged upon the minds of RIEMANN and BELTRAMI leading them to postulate as a corollary proposition to space-curvature, its manifoldness, is nothing more nor less than the intuition that the universum of spatiality cannot otherwise present itself to the intellect, owing to its peculiar adaptation to the sensuous, except by a series of continuous degrees which are perceptible only in proportion as the understanding is magnified to conform with it. After all, however, it is not improbable that the very objectivism of the universe in manifestation subsists in just the manner in which this intuitive glimpse implies and that the wisdom and utilitariness of the kosmogonic process which engendered spatiality are clearly demonstrated in that arrangement of the contents of the kosmos which presents the grossest elements of phenomena first to the intellect in its most impotent state while reserving the less crass for that time when the Thinker shall have evolved a cognitive organ adaptable to its presentations. Those metageometricians who cling to the idea of the manifoldness of space, based as we have shown upon the pseudo-interpretation of a rather vague hint arising out of an unquestionably true intuition, have allowed themselves to fall into the unconscious error of magnifying the importance of the mere insinuation as to the space-nature to such an extent as wholly to obscure in their own minds and in the minds of those who think after them whatever of the true vision that may have been grasped by them. Furthermore, it is indubitably true that that same peculiarity of arrangement by which impalpable and invisible forces really subtend gross matter producing that subtle schematism in virtue of which the visible is subjoined to the invisible, the sensuous to the non-sensuous, spirit to matter, etc., also characterizes the appearance of spatiality to the human understanding. While there is a superficial semblance of separate and discrete manifolds into which space may be divided there are, in reality, no such sharp lines of demarcation between the subtle and the gross, between the visible and the invisible or between spirit and matter, each of these being capable of reduction, by insensible degrees, into the other regardless as to whether the reductional process originates on the side of the most refined or on that of the grossest. Accordingly, there are no reasonable grounds upon which the notion of a space-manifold may be justified except as a metageometrical near-truth.

In addition to the foregoing, there are yet other very fundamental considerations which would seem to debar the totality of analytical conclusions as to the nature of space from any claim to ultimate reliability and trustworthiness. These are *first*: the fact that analyses are absolutely incapable of dealing with life; that being the direct product of a sort of mechanical consistency which marks the intellectual operations it has adaptability only for dealing with fragments or disconnected parts, and that without any reference whatsoever to the current of life or the flow of reality which has produced the parts. This fact is clearly shown in that attitude of the understanding which inevitably leads it to the declaration that a line is an infinite series of points, a plane an infinite series of lines, and a cube, an infinite series of planes, and so on, indefinitely. To do this, to look upon all phenomena as a series of parts similar to each other and piled, one upon the other, or juxtaposed in the manner which they are discovered in the sensible world, is the natural tendency of the intellect and this tendency finds its most facile expression in analytics. Inadaptability of this sort is especially observable in all problems of arithmetical analysis in which the vital element is a factor. When these analyses are carried to their logical conclusion, as has been shown in the chapter on "The Fourth Dimension," invariably they end in an evident absurdity. But it is at their very conclusion where the life-element is encountered, where reality is approached, that they break down. The failure of analysis, then, to encompass life, to fit into its requirements and to satisfy its natural outcome seems clearly to establish the basis of the perisophical nature of the entirety of analytical claims, especially that species of analysis which seeks the remoter fields of the conceptual for its determinations. *Second*: the close connection which has been seen to subsist between space and life as joint products of the same movement makes it obvious that the same ultimate rule of interpretation must be applied to both in order to insure correct and dependable judgments regarding them. How different would be the intellectual attitude towards space if it were considered in the same light as vitality, provided one really understood anything about vitality! Moreover, as it appears certain that the path of the intellect does not run in the same direction as the path which life makes, but in an inverse direction, it is clear that the judgments of the former, as to the action and essence of the latter, must necessarily be ultimately unreliable. It can readily be seen, however, that should the intellect be focused so as to follow the path of life, to attach itself to the very stream of life, it would have necessarily to neglect materiality. And such an adjustment would, of course, obviate the need of a material life at all for humanity. In fact, a physical life with an intellect would be impossible under such conditions. It is well to recognize the suitability of the present schematism and not to become unwisely restive because of it; but it is also fitting that we should discriminate between that which is possible for the intellect chained to materiality and that which is impossible for it, in such a state, when foraying in a territory foreign to its nature, and beyond its powers to master.

The predominating tendency in the intellect to account for the universe of life, mind and matter upon a strictly mechanical basis is undoubtedly due to the constitution of the intellect which does not admit of its direct consideration of the vital essence of things. We are bound ineluctably to the surface of things. All our knowledge is therefore superficial. We are even bound to the surface of ideas, and cannot penetrate to the interior of these realities. Our art is the reproductions of superfices; our philosophies are the husks of eternalities; our religions, the habiliments of relations, and while it cannot be doubted that this arrangement is pre-eminently the best possible one for the present stage of man's evolution, it is nevertheless worth while to note that it is this very restricted activity of the intellect which shuts out from man's consciousness those very elements about which he is most concerned when he goes into the field of philosophy in search of a solution to his unanswerable queries. But some progress most surely is made when the mind is enabled to see its plight and recognize what are the difficulties and limitations that lie in its path of ultimate attainment.

It is believed that the mechanistic, or true, character of the intellect reached its zenith in the mind of LAGRANGE when he succeeded in reducing the entirety of physics to certain mechanical laws and formulæ which he embodied in his "*Mecanique Analytique*" This work is undoubtedly the capstone of intellectual endeavors and stands as a monument which marks the culmination of the present stage of intellectual development. In thus placing the *Mecanique* at the apex of intellectual endeavors it is not thereby meant to be implied that the intellect shall not make more progress nor that other formulæ, equally as marvelous as those which LAGRANGE discovered, may be devised, nor that other laws, heretofore undreamed of, may be found; but what is maintained is the fact that while there will be growth and development these will run along other channels, perhaps in the realm of the intuitable, and not any longer, especially so notably as now, in an opposite direction against the current of life and reality; and further, that there will be a gradual turning away from mechanics to biogenetics, from diversity to unity, from the purely intellectual to the intuitional, and withal a final getting rid of the bonds of illusion, of that thralldom of mechanics, whereupon will slowly arise the obsolescence of all those disparities which may now be recognized in our knowledge and in the applications of the intellect to the data of the objective world.

Because the intellect is unsuited to deal with reality, and because of its peculiar adaptation for diversity, for multiplicity, due to its mechanistic *modus vivendi*, there has grown up a voluminous catalogue of systems of philosophy. These embody such a multitudinous array of beliefs, ideas, conceptions, theories and conjectures and constitute a movement in human thought which oscillates between the empiricist on the one hand and the transcendentalist on the other; between the idealist and the realist, leaning sometimes towards the Platonic, the Cartesian and the Kantian and at other times towards SPINOZA, ARISTOTLE, SPENCER and SOCRATES, always terminating by multiplying the number of diverse beliefs rather than unifying them that the conclusion is unavoidable that so marked a lack of unanimity is indicative of a profound mental prestriction. It was, therefore, inevitable that mathematics should fall under the same spell and brook no let nor hindrance until it had succeeded in devising several diverse systems of geometry which it has done for the mere joy of doing something, of following its instinctive aptitudes. There is no other basis for the heterogeneity of our philosophies, our mathematics, indeed our beliefs than this mechanical, and hence, radically illusionary character of the intellect in consequence of which we have had to be satisfied with mere glimpses, hints, intimations and faint glimmerings of reality, of life, and of those kosmic movements which, if we had the ability to trace them from their source outward, would lead us unerringly to a truer and deeper knowledge of those things that under the present schematism must remain for us a closed book.

The criterion of truth for us, constituted as we are and wedged in between the stream of life and its shore of materiality, must be that which relates our knowledge both to the stream and to the shore. It must be so that all predicates which purport to approach it shall exhibit a dual reference—one that relates to materiality and another that relates to vitality, and yet a third that shall combine these two relations into one. All assertions, therefore, which pertain exclusively to either of these elements—to materiality or to life—are necessarily partial, fragmentary and perisophical in nature. Mathematics, because it relates to matter and the mechanical forces set up by matter acting against matter cannot be said to agree with such a criterion; art, because it relates to snapshots or static views of matter is even more remote in its agreement; philosophy, as it has been known in the past and is known to-day, because it seeks to deal with a vitality fashioned after the image of materiality has failed when posited alongside of this criterion; and thus, the intellectual toil of millions of years has been in vain in so far as it has not succeeded even in raising a corner of the cover which hides reality from our view.

A near-truth is any variation from this standard, this norm or criterion. It may be either logical, cognitive, scientific or even metaphysical. To define: a logical truth is a predicate based upon and involving the coherency and consistency of thoughts themselves; a cognitive truth is the conformity of knowledge with so much of reality as is known; scientific truth is the conformity of thoughts to things and conditions. All of these are obviously near-truths. Then, too, a near-truth may be defined as an assertion based upon the criterion of truth but falling within the category of cognitive truths owing to insufficiency of data or vision. Such indeed are those metageometrical predicates—n-dimensionality, space-flexure, space-manifoldness and all other assertions based upon these in general and specifically. Any recognition of truth must clearly embrace both the vital and material aspects of its subject in order to be adequately inclusive, that is, it should include the causative, the sustentative, the relational and the developmental factors. These four factors are considered necessary and sufficient to determine the conformity of any view to the criterion of truth for when we are cognizant of the cause of a subject, understand the sustentative factors which keep it in existence, are conversant with its relations to other subjects and can follow its developmental variations until we come to its final status, why then, our knowledge is both sufficient and ultimate so far as that subject is concerned. Is it asking too much of mathematics or of philosophy or any system of thought that it conform to these standards or to this criterion before we shall accept it as final? Or shall we be satisfied with less than this? Let us hope not.

In the foregoing presentation stress has been placed upon the fragmentary, and therefore, illusionary character of the intellect in order to arrive at an understanding of the difficulties under which real knowledge has to be acquired and to indicate the inanity of all attempts to resolve the riddle of space by means of mathematics though regarded as the most typical exemplification of the mechanistic nature of the intellect. And further, to show that, on account of the radical incongruity which estranges life, the producer of spatiality, from the intellect which returns again to scrutinize the passage of life in its outward expressions, no hope of ever gaining the true viewpoint by means of the intellect need be entertained. But in doing so, it is deemed fitting that a note of warning should be sounded against any abortive attempts that may be made to obscure or distort the results of such a close discrimination lest the true import of the examination be lost for, if we emphasize the vanity of the intellect in the pursuit of that which it is by nature unsuited to attain we also equally stress the wise utilitarianism which limits it to the performance of the tasks assigned while at the same time reserving for the function of more highly evolved powers, and indeed, for the intuition itself, the solution of the riddle of spatiality. And if we declare the futility of the mathematical method in all endeavors aimed at unveiling the mysteries of life and mind, and of that movement which has its roots set in eternal duration from which it proceeds in an endless continuity of purpose and promise, we do also recognize that in the science of mathematics the intellect shall, as in no other method of cognition, most fully fulfill the kosmic intent of its existence; and moreover, in the pursuit thereof it shall push the frontiers of its possibilities outward until it can be said almost to be able to make disposition of life itself—at least to that point where, when the intuition shall have come into its own, the passage from the mechanics of matter to the dynamics of life, shall be comparatively easy and natural.

CHAPTER X

The Spiritualization of Matter the End of Evolution—Sequence and Design in the Evolution of Human Faculties—The Upspringing Intuition—Evidences of Supernormal Powers of Perception and the Possibility of Attainment—The Influence and Place of the Pituitary Body and the Pineal Gland in the Evolution of Additional Faculties—The Skeptical Attitude of Empirical Science and the Need for a More Liberal Posture—The General Results of Pituitarial Awakening Upon Man and the Theory of Knowledge.

Evolution is a continuous process and the primal impetus back of the great on-flowing ocean of life acts infinitively. It is not terminated when life has succeeded in perfecting a form for the perfection of forms in themselves is not the end of vital activity. *The end of evolution is the complete spiritualization of matter.* So that it does not matter how perfect a form may be either subjectively or in its adaptation to environments; it does not matter how faultless a medium for the ensouling life it may be, there is ever the eternal necessity that life must drive it back over the path of its genesis until it shall be transmuted into pure spirit. Adaptation succeeds adaptation and with each there is a change in the form and this process continues until there is a more or less perfect congruence between form and juxtaposed environmental conditions. But no sooner than agreement has been attained under one set of conditions new conditions arise and require a new setting, new adaptative movements. Thus there is a continuous proceeding from stage to stage, going from the grossest to the subtlest and most refined, always the form is being pushed onward and upward by life. But adaptation is not undergone for the benefit of the form, but more truly for the informing principle. It is the progression of the life-element which constitutes the adaptation of form to form and to their peculiar environs.

The form is a tool or instrument of life which it discards the moment it fails to respond to its requirements. Thus forms are constantly being assumed and as constantly being relinquished. But no effort of life is lost regardless as to whether the action is performed in one or another form. The totality of matter is perpetually being acted upon by the totality of life. Every appulse of life against matter means an added push in the direction of spiritualization. The totality of such appulses of life against matter may seem infinitely small in the visible results which they produce in the process of spiritualization; but with each there is an eternal gain in that movement that shall end in the complete transmutation of materiality into spirituality. This action of life in metamorphosing matter, the nether pole of the great pair of opposites, into spirituality, its copolar factor, in its outward, visible effects, is what we vaguely call evolution. And such it is; for life is merely unfolding that which it has enfolded. Matter, having been involved as a phase of kosmic involution, is now being evolved.

In the genesis of the kosmos there appear to be three great undulations in the universal current of life. The first of these prepares the field by depositing that elemental essence which is to become the world-plasm; the second precipitates the universum of materiality, spatiality and intellectuality, not as we now know them, of course, but as potencies; the third great undulation in the current of life effects the endowment of the world-plasm with those tendencies that are to build around themselves forms appropriate to their fulfillment. This ensoulment of the world-plasm with tendencies and the consequent segmentation of it into separate forms by these tendencies constitute the primary stages of that procedure of life which results ultimately in the up-raisement of matter and its final exaltation into pure spirit. Hence, the entire mass of materiality is besieged on all sides by the sum-total of life and the former is being raised slowly and irresistibly to heights that are immeasurably more sublime than its present degree of grossness.

It appears paradoxical, therefore, that life, although in all respects vastly superior to matter, should become the apparent vassal of materiality and give itself up to all the strict rules of imprisonment which are imposed upon it by the properties and qualities which we observe in matter. It seems so subject to every whim and fancy of matter that one is inclined to think that matter and not life is the chief designer of universal destiny. This is not a condition to be wondered at so much, for the reason that this apparent vassalage, this seeming enslavement of life by matter, is due to that superior and most marvelous adaptability of life which it enjoys in contradistinction to the relative unpliability of matter, and due also to the fact that life is kinetic and matter, being a mere deposit of life, is static. Life is mobility while matter is immobility and thus in possessing a greater range of freedom is, of course, correspondingly superior; but in this adaptation of itself to the labyrinthine cavities and multiformed interstices in matter it exhibits but a seeming serfdom which is really not a serfdom but a mastery. It is as if a man had taken lumber, hardware and stone and built a house wherein he might dwell—life has merely used matter, molded and fashioned it so as to make for itself a medium, a dwelling-place wherein it operates, not as a slave but as a master possessing unlimited freedom of motility. In the production of a form life stamps upon it, once for all time, the path of its engendering action. It leaves its finger-prints upon the mold which it makes for itself. So that if we would know where life *has been* or where it *is* we should look for its finger-prints (organization); we should observe the sinuosities which mark its pathway, remembering always that it is life that has formed the intricacies and complexities of the form into which it pours itself so accommodatingly in order that it may raise that form, develop and transmute it into something higher and better.

When we speak of *form* it must not be understood thereby that reference is made only to the gross physical form, but to the entire range of vital assumptions or vehicles which life ensouls for purposes of manifestation. This range we believe to cover the whole path of kosmogenesis seriating from the densest to the most subtle. Our chief concern, however, is the immediate effect which the totality of life's operations will have upon humanity or the form which it ensouls as the human organism. For it is impossible that humanity shall escape either the general or the specific results of the exalting power which life exerts over materiality and its appurtenances. It is, of course, impossible here to go into the various implications of this general forward movement of the universum of materiality or even to outline briefly the divergent lines of operation into which a satisfactory exposition of this view would naturally lead. And then to do so would be inappropriate in a volume of this kind. So we shall have to be content at this juncture to limit our study to a consideration of what we believe to be some of the immediate indications of this vast and most far-reaching phenomenon.

In the chapter on the "Genesis and Nature of Space" it is shown that the material universe is engendered at the same time and by the same movement or process as the universum of spatiality and intellectuality and that as the passage from chaos to kosmos proceeds the function of this movement is changed gradually from engenderment to exaltation wherein materiality is transmuted into spirituality. It is, of course, obvious that as materiality is exalted so are spatiality and intellectuality; and that as the one becomes more and more refined, capable of answering to higher and yet higher requirements so do all the others. For, at work in all and through all of these, is the current of life which pervades them, engendering, sustaining and elevating as it proceeds. So that as matter has evolved added characteristics and properties, each answering to a given need and arising out of the necessities inhering in the stage at which it appeared, so has the intellect evolved faculties to correspond therewith. In other words, the evolution of faculties for the expression of the human intellect has proceeded synchronously with the evolution of material qualities. And whenever a new faculty or an additional scope of motility is achieved by humanity there is always found a set of kosmic conditions which answers thereto. The cardinal principles of the doctrine of evolution are not, therefore, adverse to the conclusion that the organs of sense-perception—hearing, touch, sight, taste and smell—have not been endowed upon the human race or attained by it at one time; but rather that each answering to a newly acquired need and opening a wider scope of motility for the intellect has been evolved separately and in due order. It would also seem that the quality of consciousness, as it has been manifested in the various stages of life through which it has passed, and especially the mineral, vegetal and lower animal, has not always been of the same degree of efficiency. Nor has it enjoyed the same kind of freedom which it now enjoys in the highly evolved *genus homo*. It is equally apparent that matter itself has not always been in possession of the same qualities and characteristics which it now exhibits; but that it, too, has gone through various stages of evolution bringing forward into each new stage the transmuted results of each preceding one as a basis for further evolution and expansion. The innumerable archæological evidences which support this view make it unnecessary to do more than state the facts, as they appear to be substantiated by indubitable testimonies. Furthermore, it is believed that the outstanding implications of these phenomena will not be successfully controverted by those who are disinclined to see such implications in the evolutionary process. In a previous chapter we have briefly sketched the characteristics which mark the upspringing of a new faculty showing how, at first, it appears as an abnormality which exhibits itself in a very few individuals only, and that in a more or less indefinite manner; and how later the number of individuals in which it appears gradually increases, the definiteness of the faculty, at the same time, appearing more marked; then, like a tidal wave, it recurs in a still larger number of persons until, at last after a long period of time usually several thousands of years, it becomes universal exhibiting itself in every individual and appearing as a hereditary characteristic of the entire human race. It is, therefore, not without assurance as to the ultimate soundness of this view that we make the assertions which follow this brief introduction.

It has already been stated that for a very obvious reason, namely, the satisfaction of the needs of our present humanity, the intuition is for the time dominated by the intellect and held in subjugation by it so that all of man's external operations are governed and dictated almost entirely by the intellectuality, allowing the intuition only rare moments when it can come to the fore at all. This is the rule in the evolution of faculties and characteristics. The higher faculty, although potentially present in every way, is ever held in abeyance while the lower is brought, under the rigors of its own evolution, to a point where its joint operation with the higher may be executed with the least possible friction and retardation as also with the greatest possible coördination and coöperation. Accordingly, notwithstanding the fact that materiality must possess in potentiality all the qualities which it will at any time reveal, it is nevertheless necessary that these qualities shall come forth gradually and in due order. Similarly, humanity has come into possession of its various faculties of mind, and powers of physiological functions, by insensible degrees, the higher always being held in abeyance until the lower is fully developed. Those faculties which are to bestow added powers, additional freedom and a greater scope of motility are the ones which appear later than those which are truly primitive in character. These facts have been amply demonstrated by the science of embryology wherein it is shown that *ontogeny* is a recapitulation of *phylogeny*. That is, the history of the development of the individual is a recapitulation of the development of the species. Thus the various stages of development through which the human embryo passes while *in utero* are but a repetition of similar stages through which the entire human species has passed in its phylogenetic development. Wherefore, it is certain that humanity has not attained, at one and the same time, all the powers of mind and body which it now possesses; that the childhood of the human race represented a time when it had but few faculties or organs of sense-perception—indeed a time when the higher sense-organs of smell, taste and sight were entirely lacking although residing in potentiality therein.

It is undoubtedly true that the earth has passed through a similar evolution with respect to its own material characteristics, that its childhood was, in all points, analogous to the childhood of humanity; that the air, earth and water were wholly absent, except in potentiality, during the nebulous youth of its genesis. It is even probable that there are at work to-day processes which in the future shall culminate in the evolution of newer, higher and more complicately organized species of plants, animals and minerals. Every year brings fresh evidences that crystallize the conviction that the earth has been the scene for the appearance of many strange orders of animal life. Fossiliferous strata are continually yielding incontestable testimonies of changing flora and fauna. We count the animal and vegetal life of to-day as being more highly developed than that of any other previous age, and it is well that this is so, for simplicity of organization and primality of manifestation are always succeeded by complexity and a greater scope of adaptability.

We have said that the whole of that movement of the intellect which has brought forth the metageometrical creations of hyperspaces, the curvature of space and its manifoldness together with the entire assemblage of mathetic contrivances are merely the early evidences of the appearance in the human race of a new faculty, a new medium of perception whereby the Thinker shall acquire a still greater range of motility than that now offered by the intellect. Attention has been called also to the fact that this phenomenon has been manifested not alone in the field of mathematics, but in art, religion, politics and also in science in which we have only to witness the marvelous strides already made in the discovery of radio-active substances, the ROENTGEN, BECQUEREL, LEONARD and other kinds of rays. It Is quite confidently believed that these forward movements in every branch of intellectual pursuit, these combined efforts of the intellect, in peering into the occult side of material things, are in response to the evolutionary needs of the Thinker, and in addition, are the evidences, and shall in time be the cause, of the development of an additional set of faculties. Function, or the performance of acts, determines faculty or the power of action and ultimately the organ itself. Thus the mere wish to perform aroused by desire and vitalized by the will actually terminates, in the course of time, in the genesis of a faculty, or the power to perform. The constant upreaching yearnings of the Thinker through his intellect for greater freedom and a larger scope of action, the desire to peer into the mysteries of life and mind, the infantile out-feelings of the mentality after some safer and surer basis for its theory of knowledge cannot fail in producing not only the faculty or power to satisfy these cravings but the very organ or medium by virtue of which the satisfaction may be attained.

It is not strange that in mathematics the intellect should have found first the clue to the existence of a higher sphere of intellectual research wherein it might become the creator of the various entities which peopled the new found domain; it is not strange that the mathematician should, in this instance, have assumed the role of the prophet proclaiming by various mathetic contrivances (although unconsciously) that the human race is nearing that time when it shall actually be able to function consciously in some higher sphere; neither is it to be wondered at that the voice of the prophet is heard and respected throughout the earth; for, indeed the mathematician is a spokesman who, as a rule, is unmoved by sudden outbursts of passion and ecstatic frenzies of emotions but calmly and dispassionately verifies his conclusions, tests them for consistency and having found them to satisfy the most rigorous mathetic requirements hesitates not to propound them. For this cause humanity respects the mathematician, and when he speaks listens to his voice. It is well, too, that this is so; for the history of mathematics is clearly the history of the development of the intellect. So exact a determinator of the quality of intellectual efficiency is it that its reign may be said to be an absolute monarchism whose lines of dominance extend to the minutest desire or appetency. It has always been the guide of the intellect, going before, as it were, blazing the trail, pushing back the frontiers of the intellect's domain and clearing away the *debris* so that the intellect with its retinue of servitors might have an easy path of progress.

Mathematics, however, has not the aptitude to serve the intuition as it serves the intellect. So the path into which the intuition would lead humanity the mathematician, because of his training and peculiar functions, is unprepared to enter. It is for this reason that when mathematics leads the intellect up to that point where it encounters life it fails, it becomes confused and its dictatorship becomes a mockery, its decrees remain unexecuted and futile. In taking this view we have certainly no desire to offend the mathematician or to detract from the glory of his monarchistic rulership over the intellectual progress of the race; for, in truth, mathematics is the diadem of gold wherewith man has crowned his intellect. Yet it is well, yea imperative in the light of recent developments in the realm of hyperspace, that a careful discrimination should be made as between the sphere of the intellect and that to which the intuition shall attain.

The intuition, long held in abeyance until the intellect should be fully crowned and reach the zenith of its powers, is now coming to the front. It will be many centuries perhaps before it shall have grown to such proportions as those already attained by the intellect; perhaps a few thousand years may pass before the intuition shall have evolved to that point where it may labor as coadjutor to the intellect; but undoubtedly the time will come when it, too, shall reward the Thinker's labors with that which shall be more precious than the crown of gold which the intellect has won. Then, the intellect, grown old and decrepit with years of reigning shall become dim and crystal-shaped and finally pass into automatism or reflexive movements where without the urge of volitional impulses it will perform with exactness, precision and utter loyalty the tasks which it has learned so well to execute in the days of its forgotten glory. Mankind will then be free. A new freedom, wherein the erstwhile lightning flashes of intuition will become fused into one glorious sheen of all-revealing light, shall come to men and thus the race resplendent will walk the earth enshrined in the majesty of divine powers attained as a result of millions of years of aspiration.

That there are supersensuous realms so far above the range of our senses as to be entirely beyond their ken needs now no proof or argument; for the scientist has demonstrated, by the invention of instruments of extreme delicacy and precision, that such a world does really exist. Already we know of stars so distant that, though light traverses in the brief space of an hour six hundred million miles, they might have ceased to shine before the pyramids were built and yet be visible to us in the skies. If the human eye were as sensitive as the spectroscope many thousand tints and shades might be added to the world of color; if they possessed the magnifying powers of the microscope we should live in constant terror and awe of the monstrous entities that teem in the water which we drink and in the air which we breathe; and if our ears could detect the microphonic vibrations which register in the delicate apparatus of some microphones the dead, vacuum-stillness of nature's great silences would appear as a babel of voices by the seaside. The sense of touch, responding to the same range of vibrations as the micrometer, would reveal actually the interstices between particles of the densest elements; and gold, silver, platinum and mercury would seem but honeycombs of matter. But, to the forward-looking there is no element of absurdity in the expectation that all these senses shall, one day, be able to dispense with the artificial aid of physical apparatus and perform, with even greater precision and faithfulness, the task which they now perform so crudely and ineffectively. There are without doubt vibrations of taste and smell which are so far above the range of these senses that they have no effect upon them whatsoever. Notwithstanding the fact, however, that the galvanometer, microscope, the microphone, the spectroscope and the telescope have extended thus the sphere of sense-knowledge there are yet subtler vibrations to which these delicate instruments do not and ought not be expected to respond. But to say, as do many empiricists, that since these phenomena cannot be detected by scientific instruments they do not, therefore, exist seems to be expecting too much of material means as well as exposing oneself unnecessarily to criticism on the grounds of extreme materialistic appetences.

There is indeed need of a more liberal attitude among men of science towards the world of the unseen. Intolerance of the data which it offers will for a time perhaps preserve the aloofness of scientific dogmatism inviolate but there will most surely come a reaction against the dogmatism of science and men will seek freedom and attain it despite their fetters. Sir OLIVER LODGE, in his book, the *Survival of Man*,[29] says: "Man's outlook upon the universe is entering upon a new phase. Simultaneously with the beginning of a revolutionary increase in his powers of physical locomotion—which will soon be extended to a third dimension and no longer limited to a solid or liquid surface—his power of reciprocal mental intercourse is also in process of being enlarged; for there are signs that it will some day be no longer limited to contemporary denizens of earth, but will permit a utilization of knowledge and powers superior to his own, even to the extent of ultimately attaining trustworthy information concerning other conditions of existence."

It is the author's good fortune that he has for a period extending over several years been able to verify the conclusions which Sir OLIVER LODGE expresses in the above, and thus to satisfy his own mind that the process by which man's mental powers are "being enlarged" is indeed demonstrable by actual observation and experimental methods.

LODGE continues:

"The boundary between the two states—the known and the unknown, is still substantial, but it is wearing thin in places, and like excavators engaged in boring a tunnel from opposite ends amid the roar of water and other noises, we are beginning to hear now and again the strokes of the pick-axes of our comrades on the other side."

CAMILLE FLAMMARION[30] cites 436 cases of psychic manifestations examined by himself and which establish beyond any reasonable doubt that there are certain perceptive faculties, namely, clairvoyance and clairaudience, that crop out in certain individuals, sometimes in abnormal conditions and sometimes normally, the very unusual character of which proves their rudimentary nature and the potency of their maturescence in the humanity of the future. Among the cases cited by FLAMMARION are 186 instances of manifestations from the dying received by persons who were awake; 70 cases were manifestations received by persons asleep; 57 were observations of direct transmission of thought without the aid of sight, hearing or touch or other physical means; 49 were cases of sight at a distance or clairvoyance by persons awake, in dreams or in somnambulism and 74 cases of premonitory dreams or predictions of the future. Indeed, there are few persons now living who have not had similar experiences, if not exactly like these, of the same nature. These examples, of course, may be greatly multiplied in every country in the world, and it is unnecessary to enumerate them further; for, when once the existence of such faculties has been demonstrated in persons, either in a normal or an abnormal condition, their presence can no longer be questioned by the fair-minded. It is, then, only a question of evolution before they will appear in the normal way and their universalization, as transmissible characters, be an accomplished fact. When we are brought face to face with this sort of phenomenon which seems to be increasing rapidly the conclusion is inevitably forced upon us that since evolution must be a continuous process and matter destined to yield higher and more refined powers and humanity to come into a far more extensive scope of motility because of the opening avenues of knowledge, it is not impossible that these acuter senses, these new faculties are now existing in the human race in a rudimentary stage and are designed to become the universal possession of all. That this is to be the almost immediate outcome of the perpetual exalting power which life exercises not alone over materiality but over human organs and faculties as well, seems to be the one big, outstanding implication of the evolutionary process. The presence of such functions as the ability to sense the invisible and the inaudible, to answer to vibrations far subtler than anything in the scope of our external sense-organs, certainly indicates the existence of rudimentary faculties which make these functions possible. *Back of these vague, indefinite functions, back of every supernormal or abnormal manifestation of man's mentality and back of all that class of phenomena which take their rise out of supersensuous areas must lie, in ever increasing potency, faculties and organs, however rudimentary, which are the source of these manifestations.* Life, that ineluctable agent of creation, which is incessantly pushing outward the confines of the intellect's scope of motility, never wearying, never tiring nor sleeping, has long ago, in the dim and distant past of man's evolution, laid the foundations; and in fact, with one stroke of its creative hand, has molded the organs which are to become the active media of these new faculties. And now, these incipient demonstrations, these infantile struggles which we see now and again outputting from them, are but the specializing processes through which, in their later development, these organs are proceeding. These are the outward signs which should tell us that life is breaking up these organs into special parts, assigning to each a certain division of labor and making of each a perfect coördinate of all the others. It is, by these very dispread exhibitions, cutting up, specializing and by slow degrees determining the function, character and general tendence of the organs of expression wherewith these manifestations shall be centralized and put into effective operation. In doing this, it is but following its accustomed procedure, the procedure which it adopted when it produced the eye, the ear, the heart and the spleen. We shall, therefore, gauge our understanding of the purport and end of evolution; in fact, we shall determine our exact intellectual comprehension of life itself, by the attitude which we adopt towards it and the mode of its appearance. Much depends, accordingly, upon the posture which we assume towards life—whether we shall say the totality of life's creative powers has been dissipated in the bringing of the human body to its present degree of perfection; whether we shall say that it is neither necessary nor possible for life to produce other organs and faculties which shall respond to the unseen world about us revealing its glories in a way far more perfect than do our external sense-organs reveal the wonders of the world of sensation; or, whether we shall conclude from these most palpable evidences that life has yet other powers and faculties which it designs to bestow upon the human mind and other organs and capabilities with which it shall endow the human body so that man, in his evolution, shall be enabled to rise to still higher spheres while yet incarnate. There may be, and undoubtedly are, those who, for various reasons prefer to take the former positions and there are certainly those who like LODGE, FLAMMARION, HUDSON, CROOKES and a host of others, preferring the latter view, would rather believe in the strength of the great mass of corroborative testimonies that we are even to-day in the midst of the matutinal hours of a newer, a better and a far more efficient era of human evolution than any through which we have hitherto come.

Already, recent scientific investigations and the results obtained therefrom have begun to turn the attention of medical authorities to the activities of two very small organs situated in the mid-brain and known as the *pineal gland* and the *pituitary body*. These organs, and especially the *pineal gland* hitherto supposed to be a vestige of the past, are now beginning to be recognized as rudimentary organs belonging to the future evolution of humanity. Dr. CHARLES DE M. SAJOUS, who is an authority on the *pituitary body,* believes that it has no active internal secretions but is an "epithelio-nervous organ" which controls, through nerves leading to the adrenals and thyroid bone, the processes of general oxygenation, metabolism and nutrition. Little is known of the functions of the pineal gland except that it is an ovoid, reddish organ attached to the posterior cerebral commissure projecting downward and backward between the anterior pair of the *corpora quadrigemina.* It is otherwise known as the "*conarium*" the "*pinus*" or "*epiphysis.*" Situated at the base of the brain, it is held in position by a fold of the *pia mater* while its base is connected with the cerebrum by two pedicles. It contains amylaceous and gritty, calcareous particles constituting the brain sand. There are, however, marked structural resemblances between the *pineal gland* and the *pituitary body* and their formation is perhaps the most interesting feature of the development of the *thalamencephalon* or mid-brain. The *hypophysis cerebri* or *pituitary body* is a small, ovoid, pale, reddish mass varying in weight from five to ten grains and situated at the basal extremity of the brain in a depression of the cranium known as the *sella turcica,* a configuration very much like a Turkish saddle in shape. It is a composite, ductless gland and consists of two divisions, an anterior and a posterior, connected by an intermedial portion—all of which are attached to the base of the cerebrum by the *infundibulum.* The anterior lobe is larger than the posterior and very vascular, springing in its development from the buccal cavity of the embryo; the posterior lobe is situated in a depression of the anterior and is a brain-process. The *pituitary body* itself is lodged in a cavity of the *sphenoid* bone called the *pituitary fossa.* This is a most remarkable position, for the reason that the *sphenoid, or wedge-shaped,* bone which lies at the base of the skull articulates from behind with the occipital and in front with the *frontal* and *ethmoid* bones and by lateral processes with the *frontal, parietal* and *temporal* bones. From this position it binds together all the bones of the cranium, and moreover, articulates with many bones of the face. It is upon the upper surface of the *sphenoid* bone which occupies such a prominent and commanding position in the cranium, in a deep depression, that the *pituitary gland* is located.

Each nasal chamber is lined by a mucous membrane called the *pituitary* or *Schneiderian.* This membrane is prolonged into the meatuses and air sinuses which open into the nasal chambers. The *pituitary* membrane is thick and soft and diminishes the size of the meatuses and air sinuses. It is covered by a ciliated columnar epithelium and contains numerous racemose glands for the secretion of mucous or *pituita*. It is also vascular and the veins which ramify it have a plexiform or net-work like arrangement. It divides into two membranes—a respiratory, which is concerned in breathing, and an olfactory region. The respiratory region corresponds to the floor of the nose, to the inferior turbinated bone and to the lower third of the nasal septum. The olfactory region is the seat and distribution of the olfactory nerve and corresponds to the base of the nose, to the superior and middle turbinals and the upper two-thirds of the nasal septum.

Recent developments prove that this gland has a profound influence over the animal economy. It is believed by some that the *pituitary body* actually destroys certain substances which have a toxic influence on the nervous system; others believe that it secretes material media for the proper action of the trophic or nutritive apparatus; still others believe that it influences blood-pressure. It is known, however, from experimentation, that its removal in dogs, cats, mice and guinea pigs causes a fall of temperature, lassitude, muscular twitchings, dyspnœa or difficult breathing, and even speedy death. Hypertrophy of the gland is directly associated with certain diseases, such as *giantism* and *acromegaly*. The latter is a disease which causes a general enlargement of the bones of the head, feet and hands, usually occurring between the ages of twenty and forty years, and most frequently in females. The fact that these diseases are so closely associated with a hypertrophic condition of the pituitary gland has led to the conclusion that perhaps the giants or Cyclops of ancient times were cases of *giantism* or *acromegaly*. This view, while interesting from the standpoint of the functions of the *pituitary* gland, is not necessarily a correct one; for the age of giants, when men attained to a much larger stature than at present, can be accounted for on other grounds, namely; that the early mesozoic man, on account of his having to live among animals, trees and other vegetation of such huge size, had naturally to be fitted with a frame proportional to other animals in order that he might successfully cope with his environing conditions. Nature thus wisely fitted him for the conditions which she had prepared for the scenes of his life.

The facts adduced in the foregoing description are purely empirical and may be verified by any who seek to establish their correctness or incorrectness. But we are about to introduce a species of testimony which while it may also be verified will not be found so easy of verification as the above-mentioned physiological facts, and not by the same means; yet they are nevertheless deserving of a place here. It is the liberal attitude that we must adopt towards all phenomena, excluding none that give promise of the widening and deepening of our knowledge and an explanation of much that has seemed heretofore unaccountable.

We have noted how subtle is the physical connection between these two bodies, the *pineal gland* and the *pituitary body*; we have seen how profound is the effect which the latter has been demonstrated, in a measure, to have over the entire bodily economy; but there is even other testimony to the effect that those gifted with the inner vision can observe the "pulsating aura" in each body, a movement which is not unlike the pulsations of the heart and which never ceases throughout life. In the development of clairvoyance it is known that this motion becomes intensified, the auric vibrations becoming stronger and more pronounced. The *pituitary body* is the *energizer* of the *pineal gland* and, as its pulsating arc rises more and more until it contacts the *pineal gland*, it awakens and arouses it into a renewed activity in much the same manner as current electricity excites nervous tissue. When the *pineal gland* is thus aroused clairvoyant perception is said to become possible. These are facts which cannot be proved by the materialistic man of science nor can they be demonstrated to the layman who has to depend alone upon sense-deliveries for his knowledge. This is true for the reason that, in the first place, it is necessary that he shall either feel in his own mid-brain the energizing activity of these two organs and have his entire nerve-body shaken from crown to toe by the down rushing currents of that subtle energy with which the *pituitary body* floods it or be himself the perceiver of its activities. Nevertheless attention is here called to these phenomena and the conclusions drawn therefrom are offered as a means of denoting the probable line of investigations which will establish the directions which we should pursue and the source whence we shall find outcropping the new faculties and their organs of expression.

We confess to a knowledge of the fact that men of empirical science have long maintained a rather skeptical, if not contemptuous, attitude towards all these phenomena but it is also felt that there is far more of discredit in their attitude than of credit; for, in so doing, they have voluntarily adopted measures by means of which the knowledge that they so eagerly seek is shut out from their attainment. In vain, then, is appeal made to the intellect to remove the barriers which it unconsciously interposes between itself and the goal of its pursuit; in vain do we appeal to the materialist to give ear to testimony the data of which cannot be made amenable to his knife and scalpel neither to the microscope nor microphone; in sheer vanity is he adjured to look *within*—into the interior of life, of mind and the things which he handles with his instruments—for the answers to his queries, for the path which leads into the wake of life and consciousness. Because his utter loyalty and devotion to the *modus vivendi* of the intellect will not permit this; but, after all, it is not wholly wise to allure him away unbetimes from his search after truth through superficialities nor to inveigle him into giving up his tenacious prosecution of the physically determinable. We would not have it so; for, perchance, he, too, one fine day, in the quiet of his laboratory shall come upon the data which may substantiate in his own mind the long settled conclusions of the occultist who, frequently and not without cause, grows impatient at the scientist's obstinate delay. These two workers, the empiricist and the occultist, must ultimately come together as collaborators—the one working upon the form, the vehicle, physical matter and the other seeking to understand the life, the interior forces which produce, the creative element. They cannot remain always aloof from one another; for they, too, are as men digging a tunnel from opposite ends. Finally, the partition will break and thus will dawn a new day for the knowledge of humanity and men will see the rationale, the truth and good sense of coöperation in this respect.

It can be said with confidence that whatever in the future may be learned as to the physiologic functions of the *pituitary body* and the *pineal gland*, it suffices to know that it is life which they express and that, too, in a far superior manner than any of the other sense organs. The *modus* of these two glands differs in a very marked way from that of the organs of sight, hearing, taste, smell and feeling. For these latter are designed for contact with the external, objective world of sensations, their growth and evolution being dependent upon stimuli received from without while with the former the case is far different, in fact, just the opposite. The mode of life of the *pituitary body* and the *pineal gland*, instead of receiving sustenance and impetus from external stimuli, is rather dependent upon impacts received from the Thinker's own consciousness and made to impinge upon them by an exclusively interior process. Situated in the mid-brain, safely secluded from all external interference, they are naturally limited to stimuli which come from within, or it may be said, they are responsive to excitations that are more spiritual than those which come through the external sense-organs. If, as has been said they control the internal processes of metabolism (anabolism and katabolism), oxygenation, nutrition, and other important internal movements, none of which can be said to be under the control of the intellect, is it not, therefore, justly assumed that their response is directed towards stimuli which arise interiorly or upon a plane higher than the intellectual? It is a matter of scientific knowledge that those persons gifted with clairvoyance, and commonly known as "sensitives" are far more responsive to nervous excitation than those who are not so gifted. This would seem to imply that, on account of the superactivity of these two organs, the entire nerve-body has, in consequence, become more delicately and subtly organized. They seem to act as a switchboard for the regulation of the flow of the current of life through the body. Not only do they come more nearly to an adequate expression of the physiologic function of life, but, as their energization means an enlargement of the scope of perception by giving the Thinker's active consciousness access to hitherto unapproachable realities and by penetrating the outer mask which life ensouls and also laying bare a domain of unlimited knowledge the manifestation of which is far more real than anything the senses can disclose, it is evident that they constitute, in their collaborative functions, a more highly adaptable medium for the expression of the Thinker's consciousness. And if so, for the kosmic consciousness which is the source of all forms of consciousness, they furnish a specializing and *adaptizing* agency.

Now, in all those cases of inspirations, revelations, telepathic communications, clairaudience, clairvoyance, dreams, visions, etc., wherein the Thinker is enabled to perceive facts and verities which are then presented to his consciousness in a manner clearly without the province of the common sense-organs, it must be apparent that these manifestations are apprehended by a perceptual mechanism which is entirely independent of external sense presentations but which is an interior and subtler form of psychic activity. Sounds which are heard by so-called "sensitives" and objects which are perceived by eyes that are keener than those organs said to have been evolved from the "medusa" cannot be heard by other persons nor perceived by them in any way. Thus it would seem that there are inner organs of perception which respond to these finer vibrations and which enable the person so gifted to apprehend them.

There are those who, presumably basing their assertions upon actual observation and knowledge, unqualifiedly assert that in order "to gain contact with the inner worlds all that remains to be done is the awakening of the *pituitary body* and the *pineal gland*. When this is accomplished man will again possess the faculty of perception in the higher worlds, but on a grander scale than formerly (when humanity was in its infancy and exercised a lower form of psychic power only); because it will be in connection with the voluntary nervous system, and therefore, under the control of the will. Through this inner perceptive faculty all avenues of knowledge will be opened to him and he will have at his service a means of acquiring information compared with which all other methods of investigation are but child's play."[31] It is the lack of this ability to see, with our physical eyes, as it were, by the "Roentgen rays," to penetrate the inwardness of things that has baffled and confounded men for so long a time and which has eventually led certain mathematicians and others to conjecture such strange, and in many cases, illogical possibilities for the denizens of four-space. This inability together with the desire to fathom the innermost complexities of solids and to handle, albeit with unholy hands, the supersensuous, the mysterious and the unapproachable identity of "things-in-themselves" have induced the more zealous among them to contrive some kind of hypothesis which would, at least, offer an explanation of these phenomena. It has driven them to wrestle with metaphysical possibilities in a vain endeavor to grasp that which, *ignis-fatuus* like, ever evades their slightest intellectual approach. But why this prolonged struggle, why this intellectual maneuvering and sophistry? "We can calculate, compute, excogitate," says PAUL CARUS,[32] "and describe all the characteristics of four-dimensional space, so long as we remain in the realm of abstract thought and do not venture to make use of our motility and execute our plans in an actualized construction of motion; but as soon as we make an *a priori* construction of the scope of our motility, we find out the incompatibility of the whole scheme." Thus mathematicians are forced to relinquish all hopes of transforming the world of life into a sort of four-space dwelling place where everything is done according to the laws of mathematics. But whether they shall accept it or not there is a wider, truer and more rational view which recognizes all metageometrical investigations, as well as all kindred phenomena, as universal evidences indeed, as the very causes which, in the future humanity, will actually awaken and cause to be accelerated in their development these little inner sense-organs, the *pineal gland* and the *pituitary body*, whose perfect development promises to provide for the Thinker's consciousness an avenue of expression such as humanity has possessed never before. And too, it is not without full knowledge of the fact that it has been customary, among certain scientists or perhaps all of them, to regard these bodies, at least the *pineal gland*, as vestigal organs belonging to the past of human evolution, that we make these assertions. Yet, as man proceeds in the perfection of mechanical science, in the development of instruments of precision that aid his external senses, responds more and more to the subtle vibrations teeming everywhere in the atmosphere about him, and comes, in the course of time, naturally to possess a more sensitively keyed nervous mechanism, a finer body and higher spiritual aspirations, there will be a corresponding widening of his scope of vision and the attainment of larger powers of perception which must inevitably, in the very nature of things, tend towards a deeper and truer knowledge. In view of the foregoing, it is believed that the general results of this pituitarial awakening which may be expected as humanity continues to evolve should be seen in the marked effects which will be wrought in the entire metabolistic area of the human body whereby a gradual intensification and sensitization of the whole neural mechanism will raise the peculiar efficiency of all the senses whether purely physiologic or psychic. For there are undoubtedly notes so delicate in their intensity that they transcend the grasp of the audital nerves; scents and fragrances so subtle in their excelling purity that it is beyond the powers of our present olfactorial contrivances to detect them; colors and other external stimuli so sublimely supersensuous that a nervous mechanism perhaps ten-fold more delicate and responsive than ours is required to apprehend them. All these, and more than at present is conceivable, will come, with the aid of pituitarial stimulation, within the purview of a more highly developed humanity of the future. And because mathematics have led a movement into the very camp of the intellectuals—logic-bound and tethered by the severest rigors of mathesis—whereby the intolerant intellect has been compelled, by rules of its own making, to recognize the existence of the supersensuous, and by looking into the glaring light of the sun of the intuitable to gain strength of vision and boldness to press forward, a great and far-reaching service has been wrought for humanity. And in the tower of hyperspace mathematics have erected a monument to the intellect which, as long as the human race remains, will mark the great turning point in man's path to the highest life.

What if it were possible that the scientist, when he had carried instruments to their utmost precision and penetration, should suddenly, or otherwise, be endowed with a clear-perceptivity of sight, hearing and smell, so that he could with his own powers of vision, feeling and hearing take up the task where the microscope, the microphone and the micrometer left off and delve into depths far too unfathomable for his appliances, perceiving the innermost realities of things and processes? What if it were possible for him, with these added powers, to see and examine without the aid of the magnifying lens the electron, the atom and the molecule? What if the cell, the bacterium, and other invisible forms of life would then deliver up their secrets to his knowing mind? What if he could sense with his own inner vision, the ultra-violet and the infra-red rays; what indeed, if spirit itself, the innermost sheath of life, should be visible and palpable to him and he could note the internal processes, the action and movements of the infinitesimals of life? Think you not that such direct contact, such immediate and incontrovertible knowledge would be far superior to any advantage which his manufactured devices now bestow? It is even so.

Thus will react upon man's perceptive apparatus the flood of light which the awakened intuition will shed upon them and thus will man rise higher, driven on by the current of life with the mass of materiality, to a point of complete spiritualization and take additional steps in that direction which leads to Raja Yoga or the Royal Union with the divine life of the universe.

Before this step is taken, however, and before the passage from mechanics to biogenetics is made, as made it must be, man must win a complete mastery over matter. But this he will do; for more and more he is learning to put all those forms of labor which are so exacting as to leave him no time for the development of his higher powers into the hands of machinery. He will not be free until he has done this well-nigh completely. This is the task of the intellect and with it man must win his way to these higher faculties which are destined to succeed the intellect whereupon he will be ushered out of a life bound and restricted by mechanics to a life of unimaginable freedom, the *intuitive life.*

The outcome of these new faculties of perception and the development of the intuition will be the springing up of a new species of art that, turning away from appearances and sinking beneath or rising above, superficialities, will seek to portray in newly found colors, the plastic essence of things so that we shall have an art which pertains to the real, superseding that which pertains to the phenomenal. Language and the need of it will pass away; for man will have outgrown the use of signs and symbols in his communion with his fellows and will use the language of the intuition—direct and instantaneous cognition. Philosophy will be regenerated, re-created. Speculation will give way to truth and there shall be but one philosophy and that shall be the *knowledge of the real.* Mathematics, the royal insignia of the intellectual life, because it can deal only with immobilities, with segments and parts and has no aptitude for the continuous flow, will yield its kingdom to a higher form of kinetics which will serve the intuitive faculty as mathematics now serve the intellect. Science will then be no longer empirical in its method; but a system of direct and incontrovertible truths. Religion will rise to meet these changes which will come in the path of human evolution; and faith will surrender its place to knowledge. Ethics, recast in a new mold, will deal with the new aspect of man's relation to his fellowmen. Man, for whose highest good these ultimate changes will come, will be a new creature, a higher and better man; and humanity shall evolve a new race. There shall, indeed, be "a new heaven and a new earth."

THE END.